MyNotes with Integrated Review Worksheets

CALLIE DANIELS

CHRISTINE VERITY • BEVERLY FUSFIELD

MyMathLab for Essentials of College Algebra with Integrated Review

Margaret L. Lial
American River College

John Hornsby
University of New Orleans

David I. Schneider
University of Maryland

Callie J. Daniels
St. Charles Community College

PEARSON

Boston Columbus Indianapolis New York San Francisco Upper Saddle River
Amsterdam Cape Town Dubai London Madrid Milan Munich Paris Montreal Toronto
Delhi Mexico City São Paulo Sydney Hong Kong Seoul Singapore Taipei Tokyo

The author and publisher of this book have used their best efforts in preparing this book. These efforts include the development, research, and testing of the theories and programs to determine their effectiveness. The author and publisher make no warranty of any kind, expressed or implied, with regard to these programs or the documentation contained in this book. The author and publisher shall not be liable in any event for incidental or consequential damages in connection with, or arising out of, the furnishing, performance, or use of these programs.

Reproduced by Pearson from electronic files supplied by the author.

Copyright © 2015 Pearson Education, Inc.
Publishing as Pearson, 75 Arlington Street, Boston, MA 02116.

All rights reserved. No part of this publication may be reproduced, stored in a retrieval system, or transmitted, in any form or by any means, electronic, mechanical, photocopying, recording, or otherwise, without the prior written permission of the publisher. Printed in the United States of America.

ISBN-13: 978-0-321-97427-3
ISBN-10: 0-321-97427-1

6 16

www.pearsonhighered.com

PEARSON

CONTENTS

MyNotes

Chapter R	Review of Basic Concepts	**MN-1**
Chapter 1	Equations and Inequalities	**MN-57**
Chapter 2	Graphs and Functions	**MN-98**
Chapter 3	Polynomial and Rational Functions	**MN-147**
Chapter 4	Inverse, Exponential, and Logarithmic Functions	**MN-183**
Chapter 5	Systems and Matrices	**MN-226**

Integrated Review Worksheets

Chapter 1 Equations and Inequalities **IRW-1**

- 1.1R Properties of Real Numbers, Evaluating Exponential Expressions, Order of Operations, Simplifying Expressions, Operations on Real Numbers
- 1.2R Operations with Decimals; Translating Word Phrases into Algebraic Expressions and Equations; Formulas and Percent
- 1.3R Radical Notation; Simplifying Square Root Radicals; Rationalizing Square Root Denominators; Addition, Subtraction, and Multiplication of Square Root Radicals; Product Rule for Exponents; Zero Exponent Rule; Negative Exponent Rule; Add, Subtract, and Multiply Polynomials
- 1.4R, 1.5R Factoring Out the Greatest Common Factor; Factoring Trinomials; Factoring Binomials
- 1.6R Rational Expressions; Lowest Terms of a Rational Expression; Operations with Rational Expressions; Factoring (including by substitution); Negative and Rational Exponents; Simplifying Radicals with Index Greater than 2
- 1.7R Order on the Number Line; Sets and Set Operations
- 1.8R Definition and Properties of Absolute Value; Evaluating Absolute Value Expressions

Chapter 2 Graphs and Functions **IRW-111**

- 2.1R Evaluating Expressions for Given Values
- 2.2R Squaring a Binomial; Factoring Perfect Square Trinomials
- 2.3R Solving for a Specified Variable
- 2.4R Division Involving Zero
- 2.8R Operations with Polynomials; Operations with Rational Expressions

Chapter 3   Polynomial and Rational Functions **IRW-125**

    3.1R   Squaring a Binomial; Factoring Perfect Square Trinomials
    3.2R   Dividing Polynomials
    3.6R   Using Geometric Formulas; Ratio and Proportion

Chapter 4   Inverse, Exponential, and Logarithmic Functions **IRW-148**

    4.1R   Simplifying Complex Fractions
    4.5R   Rounding Whole Numbers and Decimals

Chapter 5   Systems and Matrices... **IRW-161**

    5.1R   Evaluating Expressions for Given Values; Inverse Property of Multiplication; Inverse Property of Addition
    5.8R   Identity and Inverse Properties of Addition and Multiplication of Real Numbers

Answers to Integrated Review Worksheets

    Chapter 1   Equations and Inequalities **IRWA-1**

    Chapter 2   Graphs and Functions..**IRWA-11**

    Chapter 3   Polynomial and Rational Functions **IRWA-13**

    Chapter 4   Inverse, Exponential, and Logarithmic Functions **IRWA-15**

Chapter R Review of Basic Concepts

R.1 Sets
■ Basic Definitions ■ Operations on Sets

Key Terms: set, elements (members), infinite set, finite set, Venn diagram, disjoint sets

Basic Definitions

EXAMPLE 1 Using Set Notation and Terminology
Identify each set as *finite* or *infinite*. Then determine whether 10 is an element of the set.

(a) $\{7, 8, 9, ..., 14\}$

(b) $\left\{1, \dfrac{1}{4}, \dfrac{1}{16}, \dfrac{1}{64}, ...\right\}$

(c) $\{x \mid x \text{ is a fraction between 1 and 2}\}$

(d) $\{x \mid x \text{ is a natural number between 9 and 11}\}$

EXAMPLE 2 Listing the Elements of a Set
Use set notation, and write the elements belonging to each set.

(a) $\{x \mid x \text{ is a natural number less than } 5\}$

(b) $\{x \mid x \text{ is a natural number greater than } 7 \text{ and less than } 14\}$

EXAMPLE 3 Examining Subset Relationships
Let $U = \{1, 3, 5, 7, 9, 11, 13\}$, $A = \{1, 3, 5, 7, 9, 11\}$, $B = \{1, 3, 7, 9\}$, $C = \{3, 9, 11\}$, and $D = \{1, 9\}$. Determine whether each statement is *true* or *false*.

(a) $D \subseteq B$

(b) $B \subseteq D$

(c) $C \nsubseteq A$

(d) $U = A$

Operations on Sets

EXAMPLE 4 Finding the Complement of a Set

Let $U = \{1, 2, 3, 4, 5, 6, 7\}$, $A = \{1, 3, 5, 7\}$, and $B = \{3, 4, 6\}$. Find each set.

(a) A'

(b) B'

(c) $\varnothing'$

(d) U'

Reflect: Given any set A, describe what is meant by the complement of A. What set must be defined first before determining the complement of A?

EXAMPLE 5 Finding the Intersection of Two Sets

Find each of the following.

(a) $\{9, 15, 25, 36\} \cap \{15, 20, 25, 30, 35\}$

(b) $\{2, 3, 4, 5, 6\} \cap \{1, 2, 3, 4\}$

(c) $\{1, 3, 5\} \cap \{2, 4, 6\}$

EXAMPLE 6 Finding the Union of Two Sets
Find each of the following.

(a) $\{1, 2, 5, 9, 14\} \cup \{1, 3, 4, 8\}$

(b) $\{1, 3, 5, 7\} \cup \{2, 4, 6\}$

(c) $\{1, 3, 5, 7, ...\} \cup \{2, 4, 6, ...\}$

Reflect: Discuss the difference between the union and the intersection of two sets.

Set Operations
Let A and B be sets, with universal set U.

The **complement** of set A is the set A' of all elements in the universal set that _____ belong to set A.

$$A' = \{x \mid x \in U,\ x \underline{} A\}$$

The **intersection** of sets A and B, written $A \cap B$, is made up of all the elements belonging to both set A _____ set B.

$$A \cap B = \{x \mid x \in A \underline{} x \in B\}$$

The **union** of sets A and B, written $A \cup B$, is made up of all the elements belonging to set A _____ to set B.

$$A \cup B = \{x \mid x \in A \underline{} x \in B\}$$

R.2 Real Numbers and Their Properties
■ Sets of Numbers and the Number Line ■ Exponents ■ Order of Operations
■ Properties of Real Numbers ■ Order on the Number Line ■ Absolute Value

Key Terms: number line, coordinate system, coordinate, power or exponential expression (exponential), exponent, base, absolute value

Sets of Numbers and the Number Line

Sets of Numbers

Set	Description
Natural Numbers	$\{1, 2, 3, 4, ...\}$
Whole Numbers	$\{0, 1, 2, 3, 4, ...\}$
Integers	$\{..., -3, -2, -1, 0, 1, 2, 3, ...\}$
Rational numbers	$\left\{\dfrac{p}{q} \;\middle|\; p \text{ and } q \text{ are integers and } q \neq 0\right\}$
Irrational numbers	$\{x \mid x \text{ is real but not rational}\}$
Real numbers	$\{x \mid x \text{ corresponds to a point on a number line}\}$

EXAMPLE 1 Identifying Sets of Numbers

Let $A = \left\{-8, -6, -\dfrac{12}{4}, -\dfrac{3}{4}, 0, \dfrac{3}{8}, \dfrac{1}{2}, 1, \sqrt{2}, \sqrt{5}, 6\right\}$. List the elements from A that belong to each set.

(a) Natural numbers

(b) Whole numbers

(c) Integers

(d) Rational numbers

(e) Irrational numbers

(f) Real numbers

Reflect: *Describe the set of rational numbers and the set of irrational numbers. What is the intersection of the two sets? What is their union?*

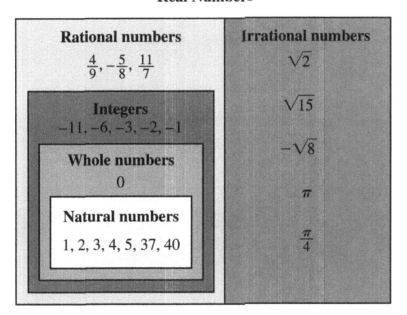

Exponents

Exponential Notation

If n is any positive integer and a is any real number, then the nth power of a is written using exponential notation as follows.

$$a^n = \underbrace{a \cdot a \cdot a \cdot \ldots \cdot a}_{\text{\_\_\_ factors of } a}$$

EXAMPLE 2 Evaluating Exponential Expressions
Evaluate each exponential expression, and identify the base and the exponent.

(a) 4^3

(b) $(-6)^2$

(c) -6^2

(d) $4 \cdot 3^2$

(e) $(4 \cdot 3)^2$

Order of Operations

Order of Operations

If grouping symbols such as parentheses, square brackets, absolute value bars, or fraction bars are present, begin as follows.

Step 1 Work separately above and below each **fraction bar**.
Step 2 Use the rules below within each set of **parentheses** or **square brackets.** Start with the innermost set and work outward.

If no grouping symbols are present, follow these steps.

Step 1 Simplify all **powers** and _____. *Work from left to right.*
Step 2 Do any **multiplications** or _____ in order. *Work from left to right.*
Step 3 Do any **negations, additions,** or _____ in order. *Work from left to right.*

EXAMPLE 3 Using Order of Operations
Evaluate each expression.

(a) $6 \div 3 + 2^3 \cdot 5$

(b) $(8+6) \div 7 \cdot 3 - 6$

(c) $\dfrac{4+3^2}{6-5 \cdot 3}$

(d) $\dfrac{-(-3)^3 + (-5)}{2(-8) - 5(3)}$

Reflect: What should you look for first when applying the order of operations?

EXAMPLE 4 Using Order of Operations
Evaluate each expression for $x = -2$, $y = 5$, and $z = -3$.

(a) $-4x^2 - 7y + 4z$

(b) $\dfrac{2(x-5)^2 + 4y}{z+4}$

(c) $\dfrac{\dfrac{x}{2} - \dfrac{y}{5}}{\dfrac{3z}{9} + \dfrac{8y}{5}}$

Properties of Real Numbers

Properties of Real Numbers

Let a, b, and c represent real numbers.

Property	Description
Closure Properties $a+b$ is a real number ab is a real number	The sum or product of two real numbers is a _____.
Commutative Properties $a+b = b+a$ $ab = ba$	The sum or product of two real numbers is the _____ regardless of their order.
Associative Properties $(a+b)+c = a+(b+c)$ $(ab)c = a(bc)$	The sum or product of three real numbers is the _____ no matter which two are added or multiplied first.
Identity Properties There exists a unique real number 0 such that $a+0 = a$ and $0+a = a.$ There exists a unique real number 1 such that $a \cdot 1 = a$ and $1 \cdot a = a.$	The sum of a real number and ___ is that real number, and the product of a real number and ___ is that real number.
Inverse Properties There exists a unique real number $-a$ such that $a+(-a) = 0$ and $-a+a = 0.$ If $a \neq 0$, there exists a unique real number $\dfrac{1}{a}$ such that $a \cdot \dfrac{1}{a} = 1$ and $\dfrac{1}{a} \cdot a = 1.$	The sum of any real number and its negative is ___, and the product of any nonzero real number and its reciprocal is ___.
Distributive Properties $a(b+c) = ab+ac$ $a(b-c) = ab-ac$	The product of a real number and the sum (or difference) of two real numbers _____ the sum (or difference) of the products of the first number and each of the other numbers.

EXAMPLE 5 Simplifying Expressions
Use the commutative and associative properties to simplify each expression.

(a) $6+(9+x)$

(b) $\dfrac{5}{8}(16y)$

(c) $-10p\left(\dfrac{6}{5}\right)$

EXAMPLE 6 Using the Distributive Property
Rewrite each expression using the distributive property and simplify, if possible.

(a) $3(x+y)$

(b) $-(m-4n)$

(c) $\dfrac{1}{3}\left(\dfrac{4}{5}m - \dfrac{3}{2}n - 27\right)$

(d) $7p + 21$

Order on the Number Line

If the real number *a* is to the left of the real number *b* on a number line, then
  a is less than b, written _____.

If the real number *a* is to the right of *b*, then
  a is greater than b, written _____.

Reflect: *Towards which number does the inequality always point?*

Absolute Value

The distance on the number line from a number to 0 is called the **absolute value** of that number.

Since distance cannot be negative, the absolute value of a number is always _____ *or 0.*

Absolute Value

Let a represent a real number.

$$|a| = \begin{cases} a & \text{if } a \geq 0 \\ -a & \text{if } a < 0 \end{cases}$$

That is, the absolute value of a positive number or 0 equals that _____, while the absolute value of a negative number equals its _____ (or opposite).

EXAMPLE 7 Evaluating Absolute Values
Evaluate each expression.

(a) $\left|-\dfrac{5}{8}\right|$

(b) $-|8|$

(c) $-|-2|$

(d) $|2x|$, for $x = \pi$

EXAMPLE 8 Measuring Blood Pressure Difference
Systolic blood pressure is the maximum pressure produced by each heartbeat. Both low blood pressure and high blood pressure may be cause for medical concern. Therefore, health care professionals are interested in a patient's "pressure difference from normal," or P_d.

If 120 is considered a normal systolic pressure, then

$P_d = |P - 120|$, where P is the patient's recorded systolic pressure.

Find P_d for a patient with a systolic pressure, P, of 113.

Properties of Absolute Value

Let a and b represent real numbers.

Property	Description
1. $\|a\| \geq 0$	The absolute value of a real number is positive or ___.
2. $\|-a\| = \|a\|$	The absolute values of a real number and its opposite are _____.
3. $\|a\| \cdot \|b\| = \|ab\|$	The product of the absolute values of two real numbers _____ the absolute value of their product.
4. $\dfrac{\|a\|}{\|b\|} = \left\|\dfrac{a}{b}\right\|$	The quotient of the absolute values of two real numbers _____ the absolute value of their quotient.
5. $\|a+b\| \leq \|a\| + \|b\|$ (the triangle inequality)	The absolute value of the sum of two real numbers _____ to the sum of their absolute values.

EXAMPLE 9 Evaluating Absolute Value Expressions

Let $x = -6$ and $y = 10$. Evaluate each expression.

(a) $|2x - 3y|$

(b) $\dfrac{2|x| - |3y|}{|xy|}$

Distance between Points on a Number Line

If P and Q are points on a number line with coordinates a and b, respectively, the distance $d(P,Q)$ between them is given by the following.

$$d(P,Q) = |b - a| \quad \text{or} \quad d(P,Q) = |a - b|$$

That is, the distance between two points on a number line is the _____ _____ of the difference between their coordinates in either order.

EXAMPLE 10 Finding the Distance between Two Points
Find the distance between -5 and 8.

R.3 Polynomials
■ Rules for Exponents ■ Polynomials ■ Addition and Subtraction ■ Multiplication
■ Division

Key Terms: algebraic expression, term, coefficient, like terms, polynomial, polynomial in x, degree of a term, degree of a polynomial, trinomial, binomial, monomial, descending order, FOIL method

Rules For Exponents

Rules for Exponents

For all positive integers m and n and all real numbers a and b, the following rules hold.

Rule	Description
Product Rule $a^m \cdot a^n = a^{m+n}$	When multiplying powers of like bases, keep the base and _____ the exponents.
Power Rule 1 $(a^m)^n = a^{mn}$	To raise a power to a power, _____ the exponents.
Power Rule 2 $(ab)^m = a^m b^m$	To raise a product to a power, raise each _____ to that power.
Power Rule 3 $\left(\dfrac{a}{b}\right)^m = \dfrac{a^m}{b^m}$ $(b \neq 0)$	To raise a quotient to a power, raise the _____ and _____ to that power.

EXAMPLE 1 Using the Product Rule
Find each product.

(a) $y^4 \cdot y^7$

(b) $(6z^5)(9z^3)(2z^2)$

EXAMPLE 2 Using the Power Rules
Simplify. Assume all variables represent nonzero real numbers.

(a) $\left(5^3\right)^2$

(b) $\left(3^4 x^2\right)^3$

(c) $\left(\dfrac{2^5}{b^4}\right)^3$

(d) $\left(\dfrac{-2m^6}{t^2 z}\right)^5$

Reflect: *What are the three power rules for expressions involving exponents?*

Zero Exponent

For any nonzero real number a, $a^0 = $ _____.

That is, any nonzero number with a zero exponent equals _____.

EXAMPLE 3 Using the Definition of a^0
Evaluate each power.

(a) 4^0

(b) $(-4)^0$

(c) -4^0

(d) $-(-4)^0$

(e) $(7r)^0$

Polynomials

EXAMPLE 4 Classifying Polynomials
The table classifies several polynomials.

Polynomial	Degree	Type
$9p^7 - 4p^3 + 8p^2$		
$29x^{11} + 8x^{15}$		
$-10r^6 s^8$		
$5a^3 b^7 - 3a^5 b^5 + 4a^2 b^9 - a^{10}$		

Reflect: How do you determine if an algebraic expression is a polynomial?

Addition and Subtraction

EXAMPLE 5 Adding and Subtracting Polynomials
Add or subtract, as indicated.

(a) $\left(2y^4 - 3y^2 + y\right) + \left(4y^4 + 7y^2 + 6y\right)$

(b) $\left(-3m^3 - 8m^2 + 4\right) - \left(m^3 + 7m^2 - 3\right)$

(c) $\left(8m^4p^5 - 9m^3p^5\right) + \left(11m^4p^5 + 15m^3p^5\right)$

(d) $4\left(x^2 - 3x + 7\right) - 5\left(2x^2 - 8x - 4\right)$

Multiplication

EXAMPLE 6 Multiplying Polynomials
Multiply $\left(3p^2 - 4p + 1\right)\left(p^3 + 2p - 8\right)$.

EXAMPLE 7 Using the FOIL Method to Multiply Two Binomials
Find each product.

(a) $(6m+1)(4m-3)$

(b) $(2x+7)(2x-7)$

(c) $r^2(3r+2)(3r-2)$

Special Products

Product of the Sum and Difference of Two Terms	$(x+y)(x-y) = x^2 - y^2$
Square of a Binomial	$(x+y)^2 = x^2 + 2xy + y^2$
	$(x-y)^2 = x^2 - 2xy + y^2$

EXAMPLE 8 Using the Special Products
Find each product.

(a) $(3p+11)(3p-11)$

(b) $(5m^3-3)(5m^3+3)$

(c) $(9k-11r^3)(9k+11r^3)$

(d) $(2m+5)^2$

(e) $(3x-7y^4)^2$

EXAMPLE 9 Multiplying More Complicated Binomials
Find each product.

(a) $[(3p-2)+5q][(3p-2)-5q]$

(b) $(x+y)^3$

(c) $(2a+b)^4$

Division

EXAMPLE 10 Dividing Polynomials
Divide $4m^3 - 8m^2 + 5m + 6$ by $2m - 1$.

EXAMPLE 11 Dividing Polynomials with Missing Terms
Divide $3x^3 - 2x^2 - 150$ by $x^2 - 4$.

R.4 Factoring Polynomials
■ Factoring Out the Greatest ■ Common Factor ■ Factoring by Grouping
■ Factoring Trinomials ■ Factoring Binomials ■ Factoring by Substitution

Key Terms: factoring, factored form, prime polynomial, factored completely, factoring by grouping

Factoring Out the Greatest Common Factor

The process of finding polynomials whose product equals a given polynomial is called _____.

EXAMPLE 1 Factoring Out the Greatest Common Factor
Factor out the greatest common factor from each polynomial.

(a) $9y^5 + y^2$

(b) $6x^2t + 8xt + 12t$

(c) $14(m+1)^3 - 28(m+1)^2 - 7(m+1)$

Factoring by Grouping

EXAMPLE 2 Factoring by Grouping
Factor each polynomial by grouping.

(a) $mp^2 + 7m + 3p^2 + 21$

(b) $2y^2 + az - 2z - ay^2$

(c) $4x^3 + 2x^2 - 2x - 1$

Factoring Trinomials

As shown here, factoring is the opposite of multiplication.

$$(2x+1)(3x-4) = 6x^2 - 5x - 4$$

EXAMPLE 3 Factoring Trinomials
Factor each trinomial, if possible.

(a) $4y^2 - 11y + 6$

(b) $6p^2 - 7p - 5$

(c) $2x^2 + 13x - 18$

(d) $16y^3 + 24y^2 - 16y$

Factoring Perfect Square Trinomials

$$x^2 + 2xy + y^2 = \underline{}$$
$$x^2 - 2xy + y^2 = \underline{}$$

EXAMPLE 4 Factoring Perfect Square Trinomials
Factor each trinomial.

(a) $16p^2 - 40pq + 25q^2$

(b) $36x^2y^2 + 84xy + 49$

Factoring Binomials

Factoring Binomials

Difference of Squares	$x^2 - y^2 = \underline{}$
Difference of Cubes	$x^3 - y^3 = \underline{}$
Sum of Cubes	$x^3 + y^3 = \underline{}$

There is no factoring pattern for a _____ *in the real number system.*

EXAMPLE 5 Factoring Differences of Squares
Factor each polynomial.

(a) $4m^2 - 9$

(b) $256k^4 - 625m^4$

(c) $(a+2b)^2 - 4c^2$

(d) $x^2 - 6x + 9 - y^4$

(e) $y^2 - x^2 + 6x - 9$

EXAMPLE 6 Factoring Sums or Differences of Cubes
Factor each polynomial.

(a) $x^3 + 27$

(b) $m^3 - 64n^3$

(c) $8q^6 + 125p^9$

Reflect: What are the special patterns in factoring?

Factoring by Substitution

EXAMPLE 7 Factoring by Substitution
Factor each polynomial.

(a) $10(2a-1)^2 - 19(2a-1) - 15$

(b) $(2a-1)^3 + 8$

(c) $6z^4 - 13z^2 - 5$

Reflect: *What is a prime polynomial? What does it mean to say a polynomial is completely factored?*

R.5 Rational Expressions
■ Rational Expressions ■ Lowest Terms of a Rational Expression
■ Multiplication and Division ■ Addition and Subtraction ■ Complex Fractions

Key Terms: rational expression, domain of a rational expression, lowest terms, complex fraction

Rational Expressions

The quotient of two polynomials P and Q, with $Q \neq 0$ is a _____.

EXAMPLE 1 Finding the Domain
Find the domain of the rational expression.
$$\frac{(x+6)(x+4)}{(x+2)(x+4)}$$

Reflect: What is a rational expression? How do you find the domain of a rational expression?

Lowest Terms of a Rational Expression

Fundamental Principle of Fractions
$$\frac{ac}{bc} = \frac{a}{b} \quad (b \neq 0, c \neq 0)$$

EXAMPLE 2 Writing Rational Expressions in Lowest Terms
Write each rational expression in lowest terms.

(a) $\dfrac{2x^2+7x-4}{5x^2+20x}$

(b) $\dfrac{6-3x}{x^2-4}$

Multiplication and Division

Multiplication and Division

For fractions $\dfrac{a}{b}$ and $\dfrac{c}{d}$ $(b \neq 0, d \neq 0)$,, the following hold.

$$\dfrac{a}{b} \cdot \dfrac{c}{d} = \dfrac{ac}{bd} \quad \text{and} \quad \dfrac{a}{b} \div \dfrac{c}{d} = \dfrac{a}{b} \cdot \dfrac{d}{c} \quad (c \neq 0)$$

That is, to find the product of two fractions, multiply their numerators to find the numerator of the product. Then multiply their denominators to find the denominator of the product. To divide two fractions, multiply the **dividend** *(the first fraction) by the* _____ *of the* **divisor** *(the second fraction).*

EXAMPLE 3 Multiplying or Dividing Rational Expressions
Multiply or divide, as indicated.

(a) $\dfrac{2y^2}{9} \cdot \dfrac{27}{8y^5}$

(b) $\dfrac{3m^2 - 2m - 8}{3m^2 + 14m + 8} \cdot \dfrac{3m + 2}{3m + 4}$

(c) $\dfrac{3p^2 + 11p - 4}{24p^3 - 8p^2} \div \dfrac{9p + 36}{24p^4 - 36p^3}$

(d) $\dfrac{x^3 - y^3}{x^2 - y^2} \cdot \dfrac{2x + 2y + xz + yz}{2x^2 + 2y^2 + zx^2 + zy^2}$

Addition and Subtraction

Addition and Subtraction

For fractions $\dfrac{a}{b}$ and $\dfrac{c}{d}$ $(b \neq 0, d \neq 0)$,, the following hold.

$$\dfrac{a}{b} + \dfrac{c}{d} = \dfrac{ad + bc}{bd} \quad \text{and} \quad \dfrac{a}{b} - \dfrac{c}{d} = \dfrac{ad - bc}{bd}$$

That is, to add (or subtract) two fractions in practice, find their least common denominator (LCD) and change each fraction to one with the LCD as denominator. The sum (or difference) of their numerators is the numerator of their sum (or difference), and the LCD is the denominator of their sum (or difference).

Finding the Least Common Denominator (LCD)

Step 1 Write each _____ as a product of prime factors.
Step 2 Form a product of all the different prime factors. Each factor should have as exponent the _____ exponent that appears on that factor.

EXAMPLE 4 Adding or Subtracting Rational Expressions
Add or subtract, as indicated.

(a) $\dfrac{5}{9x^2} + \dfrac{1}{6x}$

(b) $\dfrac{y}{y-2} + \dfrac{8}{2-y}$

(c) $\dfrac{3}{(x-1)(x+2)} - \dfrac{1}{(x+3)(x-4)}$

Complex Fractions

EXAMPLE 5 Simplifying Complex Fractions
Simplify each complex fraction. In part (b), use two methods.

(a) $\dfrac{6 - \dfrac{5}{k}}{1 + \dfrac{5}{k}}$

(b) $\dfrac{\dfrac{a}{a+1} + \dfrac{1}{a}}{\dfrac{1}{a} + \dfrac{1}{a+1}}$

Reflect: What are the two methods for simplifying complex fractions?

R.6 Rational Exponents
- Negative Exponents and the - Quotient Rule - Rational Exponents
- Complex Fractions Revisited

Negative Exponents and the Quotient Rule

Negative Exponent
Suppose that a is a nonzero real number and n is any integer.
$$a^{-n} = \frac{1}{a^n}$$

EXAMPLE 1 Using the Definition of a Negative Exponent
Evaluate each expression. In parts (d) and (e), write the expression without negative exponents. Assume all variables represent nonzero real numbers.

(a) 4^{-2}

(b) -4^{-2}

(c) $\left(\dfrac{2}{5}\right)^{-3}$

(d) $(xy)^{-3}$

(e) xy^{-3}

Reflect: *What does it mean to raise a number to the −2 power? to the −3 power?*

A negative exponent indicates _____*, not a sign change of the expression.*

Quotient Rule
Suppose that m and n are integers and a is a nonzero real number.
$$\frac{a^m}{a^n} = a^{m-n}$$

That is, when dividing powers of like bases, keep the same _____ *and subtract the* _____ *of the denominator from the* _____ *of the numerator.*

EXAMPLE 2 Using the Quotient Rule
Simplify each expression. Assume all variables represent nonzero real numbers.

(a) $\dfrac{12^5}{12^2}$

(b) $\dfrac{a^5}{a^{-8}}$

(c) $\dfrac{16m^{-9}}{12m^{11}}$

(d) $\dfrac{25r^7 z^5}{10r^6 z}$

EXAMPLE 3 Using the Rules for Exponents

Simplify each expression. Write answers without negative exponents. Assume all variables represent nonzero real numbers.

(a) $3x^{-2}\left(4^{-1}x^{-5}\right)^2$

(b) $\dfrac{12p^3q^{-1}}{8p^{-2}q}$

(c) $\dfrac{\left(3x^2\right)^{-1}\left(3x^5\right)^{-2}}{\left(3^{-1}x^{-2}\right)^2}$

Rational Exponents

The Expression $a^{1/n}$

$a^{1/n}$, n **Even**   If n is an *even* positive integer, and if $a > 0$, then $a^{1/n}$ is the positive real number whose nth power is a. That is, $\left(a^{1/n}\right)^n = a$. (In this case, $a^{1/n}$ is the principal nth root of a. See **Section R.7**.)

$a^{1/n}$, n **Odd**   If n is an *odd* positive integer, and a is *any nonzero real number*, then $a^{1/n}$ is the positive or negative real number whose nth power is a. That is, $\left(a^{1/n}\right)^n = a$.

For all positive integers n, $0^{1/n} = 0$.

EXAMPLE 4 Using the Definition of $a^{1/n}$

Evaluate each expression.

(a) $36^{1/2}$

(b) $-100^{1/2}$

(c) $-(225)^{1/2}$

(d) $625^{1/4}$

(e) $(-1296)^{1/4}$

(f) $-1296^{1/4}$

(g) $(-27)^{1/3}$

(h) $-32^{1/5}$

Section R.6 MyNotes   MN-43

Reflect: *Describe the relationship between fractional exponents and the roots of a number.*

The Expression $a^{m/n}$

Let m be any integer, n be any positive integer, and a be any real number for which $a^{1/n}$ is a real number.

$$a^{m/n} = \left(a^{1/n}\right)^m$$

EXAMPLE 5 Using the Definition of $a^{m/n}$
Evaluate each expression.

(a) $125^{2/3}$

(b) $32^{7/5}$

(c) $-81^{3/2}$

(d) $(-27)^{2/3}$

(e) $16^{-3/4}$

(f) $(-4)^{5/2}$

CHAPTER R Review of Basic Concepts

Definitions and Rules for Exponents
Suppose that r and s represent rational numbers. The results here are valid for all positive numbers a and b.

Product rule $\quad a^r \cdot a^s = $ _____ **Power rules** $\quad (a^r)^s = $ _____

$\qquad\qquad\qquad\qquad\qquad\qquad\qquad\qquad\qquad (ab)^r = $ _____

Quotient rule $\quad \dfrac{a^r}{a^s} = $ _____ $\qquad\qquad \left(\dfrac{a}{b}\right)^r = $ _____

Negative exponent $\quad a^{-r} = $ _____

EXAMPLE 6 Using the Rules for Exponents
Simplify each expression. Assume all variables represent positive real numbers.

(a) $\dfrac{27^{1/3} \cdot 27^{5/3}}{27^3}$

(b) $81^{5/4} \cdot 4^{-3/2}$

(c) $6y^{2/3} \cdot 2y^{1/2}$

(d) $\left(\dfrac{3m^{5/6}}{y^{3/4}}\right)^2 \left(\dfrac{8y^3}{m^6}\right)^{2/3}$

(e) $m^{2/3}\left(m^{7/3} + 2m^{1/3}\right)$

EXAMPLE 7 Factoring Expressions with Negative or Rational Exponents

Factor out the least power of the variable or variable expression. Assume all variables represent positive real numbers.

(a) $12x^{-2} - 8x^{-3}$

(b) $4m^{1/2} + 3m^{3/2}$

(c) $(y-2)^{-1/3} + (y-2)^{2/3}$

Complex Fractions Revisited

EXAMPLE 8 Simplifying a Fraction with Negative Exponents

Simplify $\dfrac{(x+y)^{-1}}{x^{-1}+y^{-1}}$. Write the result with only positive exponents.

R.7 Radical Expressions
■ Radical Notation ■ Simplified Radicals ■ Operations with Radicals
■ Rationalizing Denominators

Key Terms: radicand, index of a radical, principal *n*th root, like radicals, unlike radicals, rationalizing the denominator, conjugates

Radical Notation

Radical Notation for $a^{1/n}$
Suppose that a is a real number, n is a positive integer, and $a^{1/n}$ is a real number.
$$\sqrt[n]{a} = a^{1/n}$$

Radical Notation for $a^{m/n}$
Suppose that a is a real number, m is an integer, n is a positive integer, and $\sqrt[n]{a}$ is a real number.
$$a^{m/n} = \left(\sqrt[n]{a}\right)^m = \underline{\hspace{2cm}}$$

EXAMPLE 1 Evaluating Roots
Write each root using exponents and evaluate.

(a) $\sqrt[4]{16}$

(b) $-\sqrt[4]{16}$

(c) $\sqrt[5]{-32}$

(d) $\sqrt[3]{1000}$

(e) $\sqrt[6]{\dfrac{64}{729}}$

(f) $\sqrt[4]{-16}$

CHAPTER R Review of Basic Concepts

Reflect: *What conditions on n and a guarantee $\sqrt[n]{a}$ is a real number? Not a real number?*

EXAMPLE 2 Converting from Rational Exponents to Radicals

Write in radical form and simplify. Assume all variable expressions represent positive real numbers.

(a) $8^{2/3}$

(b) $(-32)^{4/5}$

(c) $-16^{3/4}$

(d) $x^{5/6}$

(e) $3x^{2/3}$

(f) $2p^{1/2}$

(g) $(3a+b)^{1/4}$

EXAMPLE 3 Converting from Radicals to Rational Exponents
Write in exponential form. Assume all variable expressions represent positive real numbers.

(a) $\sqrt[4]{x^5}$

(b) $\sqrt{3y}$

(c) $10\left(\sqrt[5]{z}\right)^2$

(d) $5\sqrt[3]{(2x^4)^7}$

(e) $\sqrt{p^2+q}$

Evaluating $\sqrt[n]{a^n}$

If n is an *even* positive integer, then $\sqrt[n]{a^n} = |a|$.

If n is an *odd* positive integer, then $\sqrt[n]{a^n} = a$.

EXAMPLE 4 Using Absolute Value to Simplify Roots
Simplify each expression.

(a) $\sqrt{p^4}$

(b) $\sqrt[4]{p^4}$

(c) $\sqrt{16m^8 r^6}$

(d) $\sqrt[6]{(-2)^6}$

(e) $\sqrt[5]{m^5}$

(f) $\sqrt{(2k+3)^2}$

(g) $\sqrt{x^2 - 4x + 4}$

Rules for Radicals

Suppose that a and b represent real numbers, and m and n represent positive integers for which the indicated roots are real numbers.

Rule	Description
Product Rule $\sqrt[n]{a} \cdot \sqrt[n]{b} = \sqrt[n]{ab}$	The product of two roots is the root of the _____.
Quotient Rule $\sqrt[n]{\dfrac{a}{b}} = \dfrac{\sqrt[n]{a}}{\sqrt[n]{b}} \quad (b \neq 0)$	The root of a quotient is the _____ of the roots.
Power Rule $\sqrt[m]{\sqrt[n]{a}} = \sqrt[mn]{a}$	The index of the root of a root is the _____ of their indexes.

EXAMPLE 5 Simplifying Radical Expressions
Simplify. Assume all variable expressions represent positive real numbers.

(a) $\sqrt{6} \cdot \sqrt{54}$

(b) $\sqrt[3]{m} \cdot \sqrt[3]{m^2}$

(c) $\sqrt{\dfrac{7}{64}}$

(d) $\sqrt[4]{\dfrac{a}{b^4}}$

(e) $\sqrt[7]{\sqrt[3]{2}}$

(f) $\sqrt[4]{\sqrt{3}}$

CHAPTER R Review of Basic Concepts

Simplified Radicals

Simplified Radicals
An expression with radicals is simplified when all of the following conditions are satisfied.

1. The radicand has no factor raised to a power greater than or equal to the index.
2. The radicand has no fractions.
3. No denominator contains a radical.
4. Exponents in the radicand and the index of the radical have greatest common factor 1.
5. All indicated operations have been performed (if possible).

EXAMPLE 6 Simplifying Radicals
Simplify each radical.

(a) $\sqrt{175}$

(b) $-3\sqrt[5]{32}$

(c) $\sqrt[3]{81x^5 y^7 z^6}$

Operations with Radicals

EXAMPLE 7 Adding and Subtracting Radicals
Add or subtract, as indicated. Assume all variables represent positive real numbers.

(a) $3\sqrt[4]{11pq} + \left(-7\sqrt[4]{11pq}\right)$

(b) $\sqrt{98x^3y} + 3x\sqrt{32xy}$

(c) $\sqrt[3]{64m^4n^5} - \sqrt[3]{-27m^{10}n^{14}}$

EXAMPLE 8 Simplifying Radicals
Simplify each radical. Assume all variables represent positive real numbers.

(a) $\sqrt[6]{3^2}$

(b) $\sqrt[6]{x^{12}y^3}$

(c) $\sqrt[9]{\sqrt{6^3}}$

EXAMPLE 9 Multiplying Radical Expressions
Find each product.

(a) $\left(\sqrt{7}-\sqrt{10}\right)\left(\sqrt{7}+\sqrt{10}\right)$

(b) $\left(\sqrt{2}+3\right)\left(\sqrt{8}-5\right)$

Rationalizing Denominators

EXAMPLE 10 Rationalizing Denominators
Rationalize each denominator.

(a) $\dfrac{4}{\sqrt{3}}$

(b) $\sqrt[4]{\dfrac{3}{5}}$

EXAMPLE 11 Simplifying Radical Expressions with Fractions
Simplify each expression. Assume all variables represent positive real numbers.

(a) $\dfrac{\sqrt[4]{xy^3}}{\sqrt[4]{x^3 y^2}}$

(b) $\sqrt[3]{\dfrac{5}{x^6}} - \sqrt[3]{\dfrac{4}{x^9}}$

EXAMPLE 12 Rationalizing a Binomial Denominator

Rationalize the denominator of $\dfrac{1}{1-\sqrt{2}}$.

Reflect: Describe your approach when rationalizing an expression with a monomial denominator. Describe your approach when rationalizing an expression with a binomial denominator.

Chapter 1 Equations and Inequalities

1.1 Linear Equations
- Basic Terminology of Equations ■ Solving Linear Equations
- Identities, Conditional Equations, and Contradictions
- Solving for a Specified Variable (Literal Equations)

Key Terms: equation, solution or root, solution set, equivalent equations, linear equation in one variable, first-degree equation, identity, conditional equation, contradiction, simple interest, literal equation, future or maturity value

Basic Terminology of Equations

A number that makes an equation a true statement is called a _____ or _____ of the equation.

The set of all numbers that satisfy an equation is called the _____ _____ of the equation.

Equations with the same solution set are _____ _____.

Addition and Multiplication Properties of Equality
Let a, b, and c represent real numbers.

$$\text{If } a = b, \text{ then } a + c = b + c.$$

That is, the same number may be added to each side of an equation without changing the solution set.

$$\text{If } a = b \text{ and } c \neq 0, \text{ then } ac = bc.$$

That is, each side of an equation may be multiplied by the same nonzero number without changing the solution set. (Multiplying each side by zero leads to $0 = 0$.)

Solving Linear Equations

Linear Equation in One Variable
A **linear equation in one variable** is an equation that can be written in the form

$$ax + b = 0,$$

where a and b are real numbers with $a \neq 0$.

EXAMPLE 1 Solving a Linear Equation
Solve $3(2x-4) = 7 - (x+5)$.

EXAMPLE 2 Solving a Linear Equation with Fractions.
Solve $\dfrac{2x+4}{3} + \dfrac{1}{2}x = \dfrac{1}{4}x - \dfrac{7}{3}$.

Identities, Conditional Equations, and Contradictions

EXAMPLE 3 Identifying Types of Equations

Determine whether each equation is an *identity*, a *conditional equation*, or a *contradiction*. Give the solution set.

(a) $-2(x+4)+3x = x-8$

(b) $5x-4 = 11$

(c) $3(3x-1) = 9x+7$

Identifying Types of Linear Equations
1. If solving a linear equation leads to a true statement such as 0 = 0, the equation is an _____. Its solution set is _____. (See Example 3(a).)
2. If solving a linear equation leads to a single solution such as $x = 3$, the equation is _____. Its solution set consists of _____. (See Example 3(b).)
3. If solving a linear equation leads to a false statement such as −3 = 7, the equation is a _____. Its solution set is _____. (See Example 3(c).)

Solving for a Specified Variable (Literal Equations)

EXAMPLE 4 Solving for a Specified Variable
Solve for the specified variable.

(a) $I = Prt$, for t

(b) $A - P = Prt$, for P

(c) $3(2x - 5a) + 4b = 4x - 2$, for x

Reflect: *How is solving a literal equation for a specified variable similar to solving an equation? How is solving a literal equation for a specified variable different from solving an equation?*

EXAMPLE 5 Applying the Simple Interest Formula

Becky Brugman borrowed $5240 for new furniture. She will pay it off in 11 months at an annual simple interest rate of 4.5%. How much interest will she pay?

1.2 Applications and Modeling with Linear Equations

■ Solving Applied Problems ■ Geometry Problems ■ Motion Problems
■ Mixture Problems ■ Modeling with Linear Equations

Key Terms: mathematical model, linear model

Solving Applied Problems

Solving an Applied Problem

Step 1   **Read** the problem carefully until you understand what is given and what is to be found.

Step 2   **Assign a variable** to represent the unknown value, using diagrams or tables as needed. Write down what the variable represents. If necessary, express any other unknown values in terms of the variable.

Step 3   **Write an equation** using the variable expressions(s).

Step 4   **Solve** the equation.

Step 5   **State the answer** to the problem Does it seem reasonable?

Step 6   **Check** the answer in the words of the original problem.

Geometry Problems

EXAMPLE 1 Finding the Dimensions of a Square

If the length of each side of a square is increased by 3 cm, the perimeter of the new square is 40 cm more than twice the length of each side of the original square. Find the dimensions of the original square.

Motion Problems

EXAMPLE 2 Solving a Motion Problem

Maria and Eduardo are traveling to a business conference. The trip takes 2 hr for Maria and 2.5 hr for Eduardo, since he lives 40 mi farther away. Eduardo travels 5 mph faster than Maria. Find their average rates.

Mixture Problems

The concentration of the final mixture must be _____ the concentrations of the two solutions making up the mixture.

EXAMPLE 3 Solving a Mixture Problem

Lisa Harmon is a chemist. She needs a 20% solution of alcohol. She has a 15% solution on hand, as well as a 30% solution. How many liters of the 15% solution should she add to 3 L of the 30% solution to obtain her 20% solution?

EXAMPLE 4 Solving an Investment Problem

An artist has sold a painting for $410,000. He needs some of the money in 6 months and the rest in 1 yr. He can get a Treasury bond for 6 months at 2.65% and one for a year at 2.91%. His broker tells him the two investments will earn a total of $8761. How much should be invested at each rate to obtain that amount of interest?

Reflect: *Why is an investment problem considered to be a mixture problem?*

Modeling with Linear Equations

EXAMPLE 5 Modeling Prevention of Indoor Pollutants

If a vented range hood removes contaminants such as carbon monoxide and nitrogen dioxide from the air at a rate of F liters of air per second, then the percent P of contaminants that are also removed from the surrounding air can be modeled by the linear equation

$$P = 1.06F + 7.18, \quad \text{where } 10 \le F \le 75.$$

What flow F must a range hood have to remove 50% of the contaminants from the air? (*Source: Proceedings of the Third International Conference on Indoor Air Quality and Climate.*)

EXAMPLE 6 Modeling Health Care Costs

The projected per capita health care expenditures in the United States, where y is in dollars, and x is years after 2000, are given by the following linear equation.

$$y = 343x + 4512$$

(*Source*: Centers for Medicare and Medicaid Services.)

(a) What were the per capita health care expenditures in the year 2010?

(b) If this model continues to describe health care expenditures, when will the per capita expenditures reach $9200?

1.3 Complex Numbers

■ Basic Concepts of Complex Numbers ■ Operations on Complex Numbers

Key Terms: imaginary unit, complex number, real part, imaginary part, pure imaginary number, nonreal complex number, standard form, complex conjugate

Basic Concepts of Complex Numbers

The Imaginary Unit i

$$i = \sqrt{-1}, \quad \text{and therefore,} \quad i^2 = -1.$$

(Note that $-i$ is also a square root of -1.)

Complex Number

If a and b are real numbers, then any number of the form $a + bi$ is a **complex number.** In the complex number $a + bi$, a is the _____ _____ and b is the _____ _____.

The Expression $\sqrt{-a}$

If $a > 0$, then
$$\sqrt{-a} = i\sqrt{a}.$$

EXAMPLE 1 Writing $\sqrt{-a}$ as $i\sqrt{a}$.

Write as the product of a real number and i, using the definition of $\sqrt{-a}$.

(a) $\sqrt{-16}$ (b) $\sqrt{-70}$ (c) $\sqrt{-48}$

Operations on Complex Numbers

When working with negative radicands, use the definition _____ before using any of the other rules for radicals.

EXAMPLE 2 Finding Products and Quotients Involving $\sqrt{-a}$

Multiply or divide, as indicated. Simplify each answer.

(a) $\sqrt{-7} \cdot \sqrt{-7}$

(b) $\sqrt{-6} \cdot \sqrt{-10}$

(c) $\dfrac{\sqrt{-20}}{\sqrt{-2}}$

(d) $\dfrac{\sqrt{-48}}{\sqrt{24}}$

EXAMPLE 3 Simplifying a Quotient Involving $\sqrt{-a}$

Write $\dfrac{-8 + \sqrt{-128}}{4}$ in standard form $a + bi$.

Addition and Subtraction of Complex Numbers

For complex numbers $a + bi$ and $c + di$,

$$(a+bi)+(c+di) = (a+c)+(b+d)i$$

and

$$(a+bi)-(c+di) = (a-c)+(b-d)i.$$

That is, to add or subtract complex numbers, _____ or _____ the real parts and _____ or _____ the imaginary parts.

EXAMPLE 4 Adding and Subtracting Complex Numbers

Find each sum or difference.

(a) $(3-4i)+(-2+6i)$

(b) $(-4+3i)-(6-7i)$

Multiplication of Complex Numbers

For complex numbers $a + bi$ and $c + di$,

$$(a+bi)(c+di) = (ac-bd)+(ad+bc)i.$$

EXAMPLE 5 Multiplying Complex Numbers

Find each product.

(a) $(2-3i)(3+4i)$

(b) $(4+3i)^2$

(c) $(6+5i)(6-5i)$

Property of Complex Conjugates
For real numbers a and b,

$$(a+bi)(a-bi) = a^2 + b^2.$$

The product of a complex number and its conjugates is always a real number.

To find the quotient of two complex numbers in standard form, we multiply both the numerator and the denominator by the _____ _____ of the denominator.

EXAMPLE 6 Dividing Complex Numbers
Write each quotient in standard form $a + bi$.

(a) $\dfrac{3+2i}{5-i}$

(b) $\dfrac{3}{i}$

CHAPTER 1 Equations and Inequalities

Powers of i can be simplified using the facts

$$i^2 = -1 \quad \text{and} \quad i^4 = \left(i^2\right)^2 = (-1)^2 = 1.$$

$i^1 =$	$i^5 =$
$i^2 =$	$i^6 =$
$i^3 =$	$i^7 =$
$i^4 =$	$i^8 =$

Powers of i cycle through the same four outcomes (i, -1, $-i$, and 1) since i^4 has the same multiplicative property as 1. Also, any power of i with an exponent that is a multiple of 4 has value 1. As with real numbers, $i^0 = 1$.

EXAMPLE 7 Simplifying Powers of i

Simplify each power of i.

(a) i^{15} (b) i^{-3}

1.4 Quadratic Equations
■ Solving a Quadratic Equation ■ Completing the Square ■ The Quadratic Formula
■ Solving for a Specified Variable ■ The Discriminant

Key Terms: quadratic equation, standard form, second-degree equation, double solution, cubic equation, discriminant

Quadratic Equation in One Variable
An equation that can be written in the form

$$ax^2 + bx + c = 0,$$

where a, b, and c are real numbers with $a \neq 0$, is a **quadratic equation.** The given form is called **standard form.**

Solving a Quadratic Equation

Zero-Factor Property
If a and b are complex numbers with $ab = 0$, then _____ or _____ or both equal zero.

EXAMPLE 1 Using the Zero-Factor Property
Solve $6x^2 + 7x = 3$.

Square Root Property

If $x^2 = k$, then $x = \sqrt{k}$ or $x = -\sqrt{k}$.

That is, the solution set of $x^2 = k$ *is*

$$\{\underline{}\}, \quad \text{which may be abbreviated} \quad \{\underline{}\}.$$

EXAMPLE 2 Using the Square Root Property
Solve each quadratic equation.

(a) $x^2 = 17$

(b) $x^2 = -25$

(c) $(x-4)^2 = 12$

Completing the Square

Solving a Quadratic Equation by Completing the Square

To solve $ax^2 + bx + c = 0$, where $a \neq 0$, by completing the square, use these steps.

Step 1 If $a \neq 1$, divide both sides of the equation by a.

Step 2 Rewrite the equation so that the constant term is alone on one side of the equality symbol.

Step 3 Square half the coefficient of x, and add this square to each side of the equation.

Step 4 Factor the resulting trinomial as a perfect square and combine like terms on the other side.

Step 5 Use the square root property to complete the solution.

EXAMPLE 3 Using Completing the Square ($a = 1$)
Solve $x^2 - 4x - 14 = 0$.

EXAMPLE 4 Using Completing the Square ($a \neq 1$)
Solve $9x^2 - 12x + 9 = 0$.

The Quadratic Formula

Quadratic Formula
The solutions of the quadratic equation $ax^2 + bx + c = 0$, where $a \neq 0$, are given by the quadratic formula.

$$x = \frac{-b \pm \sqrt{b^2 - 4ac}}{2a}$$

EXAMPLE 5 Using the Quadratic Formula (Real Solutions)
Solve $x^2 - 4x = -2$.

EXAMPLE 6 Using the Quadratic Formula (Nonreal Complex Solutions)
Solve $2x^2 = x - 4$.

EXAMPLE 7 Solving a Cubic Equation
Solve $x^3 + 8 = 0$ using factoring and the quadratic formula.

Reflect: *How will you decide which method to use to solve a quadratic equation?*

Solving for a Specified Variable

EXAMPLE 8 Solving for a Quadratic Variable in a Formula
Solve for the specified variable. Use $\pm$ when taking square roots.

(a) $\mathcal{A} = \dfrac{\pi d^2}{4}$, for d

(b) $rt^2 - st = k \; (r \neq 0)$, for t

The Discriminant

Discriminant	Number of Solutions	Type of Solutions
Positive, perfect square		
Positive, but not a perfect square		
Zero		
Negative		

EXAMPLE 9 Using the Discriminant

Determine the number of distinct solutions, and tell whether they are *rational*, *irrational*, or *nonreal complex* numbers.

(a) $5x^2 + 2x - 4 = 0$ (b) $x^2 - 10x = -25$ (c) $2x^2 - x + 1 = 0$

Reflect: How does the discriminant give you information about the number and type of solutions of a quadratic equation?

1.5 Applications and Modeling with Quadratic Equations
■ Geometry Problems ■ Using the Pythagorean Theorem
■ Height of a Projected Object ■ Modeling with Quadratic Equations

Key Terms: leg, hypotenuse

Geometry Problems

EXAMPLE 1 Solving a Problem Involving Volume
A piece of machinery produces rectangular sheets of metal such that the length is three times the width. Equal-sized squares measuring 5 in. on a side can be cut from the corners so that the resulting piece of metal can be shaped into an open box by folding up the flaps. If specifications call for the volume of the box to be 1435 in.^{3}, find the dimensions of the original piece of metal.

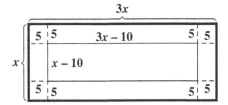
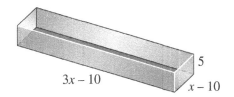

Using the Pythagorean Theorem

Pythagorean Theorem
In a right triangle, the sum of the squares of the lengths of the legs is equal to the square of the length of the hypotenuse.

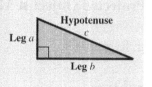

$$a^2 + b^2 = c^2$$

EXAMPLE 2 Applying the Pythagorean Theorem

A piece of property has the shape of a right triangle. The longer leg is 20 m longer than twice the length of the shorter leg. The hypotenuse is 10 m longer than the length of the longer leg. Find the lengths of the sides of the triangular lot.

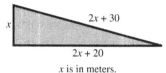

x is in meters.

Height of a Projected Object

EXAMPLE 3 Solving a Problem Involving Projectile Height

If a projectile is launched vertically upward from the ground with an initial velocity of 100 ft per sec, neglecting air resistance, its height s (in feet) above the ground t seconds after projection is given by $s = -16t^2 + 100t$.

(a) After how many seconds will it be 50 ft above the ground?

(b) How long will it take for the projectile to return to the ground?

Modeling with Quadratic Equations

EXAMPLE 4 Analyzing Trolley Ridership

The I-Ride Trolley service carries passengers along the International Drive Resort Area of Orlando, Florida. The bar graph in Figure 8 shows I-Ride Trolley ridership data in millions. The quadratic equation

$$y = -0.0072x^2 + 0.1081x + 1.619$$

models ridership from 2000 to 2010, where y represents ridership in millions and $x = 0$ represents 2000, $x = 1$ represents 2001, and so on.

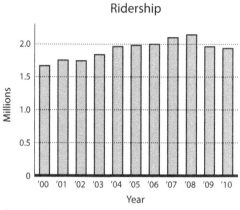

Source: I-Ride Trolley, International Drive Master Transit, www.itrolley.com

(a) Use the model to determine ridership in 2008. Compare the result to the actual ridership figure of 2.1 million.

(b) According to the model, in what year did ridership reach 1.9 million?

1.6 Other Types of Equations and Applications
■ Rational Equations ■ Work Rate Problems ■ Equations with Radicals
■ Equations with Rational Exponents ■ Equations Quadratic in Form

Key Terms: rational equation, proposed solution, equation quadratic in form

Rational Equations

A _____ _____ is an equation that has a rational expression for one or more terms.

To solve a rational equation, multiply each side by the _____ _____ _____ of the terms of the equation to eliminate fractions, and then solve the resulting equation.

Be sure to check all proposed solutions in the original equation.

EXAMPLE 1 Solving Rational Equations That Lead to Linear Equations
Solve each equation.

(a) $\dfrac{3x-1}{3} - \dfrac{2x}{x-1} = x$

(b) $\dfrac{x}{x-2} = \dfrac{2}{x-2} + 2$

EXAMPLE 2 Solving Rational Equations That Lead to Quadratic Equations

Solve each equation.

(a) $\dfrac{3x+2}{x-2} + \dfrac{1}{x} = \dfrac{-2}{x^2 - 2x}$

(b) $\dfrac{-4x}{x-1} + \dfrac{4}{x+1} = \dfrac{-8}{x^2 - 1}$

Work Rate Problems

EXAMPLE 3 Solving a Work Rate Problem

One printer can do a job twice as fast as another. Working together, both printers can do the job in 2 hr. How long would it take each printer, working alone, to do the job?

Equations with Radicals

Power Property
If P and Q are algebraic expressions, then every solution of the equation $P = Q$ is also a solution of the equation $P^n = Q^n$, for any positive integer n.

When we use the power property to solve an equation, it is essential to check all proposed solutions in the original equation.

Be very careful when using the power property. It does *not* say that the equations $P = Q$ and $P^n = Q^n$ are equivalent. It says only that _____ _____ of the original equation $P = Q$ is also a solution of the new equation $P^n = Q^n$.

Solving an Equation Involving Radicals
To solve an equation containing radicals, follow these steps.

Step 1 Isolate the radical on one side of the equation.

Step 2 Raise each side of the equation to a power that is the same as the index of the radical so that the radical is eliminated.

If the equation still contains a radical, repeat Steps 1 and 2.

Step 3 Solve the resulting equation.

Step 4 Check each proposed solution in the original equation.

EXAMPLE 4 Solving an Equation Containing a Radical (Square Root)
Solve $x - \sqrt{15 - 2x} = 0$.

EXAMPLE 5 Solving an Equation Containing Two Radicals

Solve $\sqrt{2x+3} - \sqrt{x+1} = 1$.

Reflect: Why is it incorrect to square each term individually as the first step in Example 5?

EXAMPLE 6 Solving an Equation Containing a Radical (Cube Root)

Solve $\sqrt[3]{4x^2 - 4x + 1} - \sqrt[3]{x} = 0$

Equations with Rational Exponents

An equation with a rational exponent contains a _____, or _____ _____, raised to an _____ that is a rational number.

EXAMPLE 7 Solving Equations with Rational Exponents
Solve each equation.

(a) $x^{3/5} = 27$

(b) $(x-4)^{2/3} = 16$

Equations Quadratic in Form

Equation Quadratic in Form
An equation is said to be **quadratic in form** if it can be written as

$$au^2 + bu + c = 0,$$

where $a \neq 0$ and u is some algebraic expression.

EXAMPLE 8 Solving Equations Quadratic in Form
Solve each equation.

(a) $(x+1)^{2/3} - (x+1)^{1/3} - 2 = 0$

(b) $6x^{-2} + x^{-1} = 2$

When using a substitution variable in solving an equation that is quadratic in form, do not forget the step that gives the solution in terms of the original variable.

EXAMPLE 9 Solving an Equation Quadratic in Form

Solve $12x^4 - 11x^2 + 2 = 0$.

1.7 Inequalities
- **Linear Inequalities** ■ **Three-Part Inequalities** ■ **Quadratic Inequalities**
- **Rational Inequalities**

Key Terms: inequality, linear inequality in one variable, interval, interval notation, open interval, closed interval, break-even point, quadratic inequality, strict inequality, nonstrict inequality, rational inequality

Properties of Inequality
Let a, b, and c represent real numbers.
1. If $a < b$, then $a + c < b + c$.
2. If $a < b$ and if $c > 0$, then $ac < bc$.
3. If $a < b$ and if $c < 0$, then $ac > bc$.

Always remember to _____ the direction of the inequality symbol when multiplying or dividing by a negative number.

Linear Inequalities

Linear Inequality in One Variable
A **linear inequality in one variable** is an inequality that can be written in the form
$$ax + b > 0,$$
where a and b are real numbers, with $a \neq 0$. (Any of the symbols $\geq$, $<$, and $\leq$ may also be used.)

EXAMPLE 1 Solving a Linear Inequality
Solve $-3x + 5 > -7$.

Reflect: *Describe an open interval using set notation, interval notation, and as a graph. Similarly, describe a closed interval, a disjoint interval, and the set of all real numbers.*

EXAMPLE 2 Solving a Linear Inequality
Solve $4 - 3x \leq 7 + 2x$. Give the solution set in interval notation.

EXAMPLE 3 Finding the Break-Even Point
If the revenue and cost of a certain product are given by

$$R = 4x \quad \text{and} \quad C = 2x + 1000,$$

where x is the number of units produced and sold, at what production level does R *at least equal* C?

Three-Part Inequalities

EXAMPLE 4 Solving a Three-Part Inequality
Solve $-2 < 5 + 3x < 20$.

Quadratic Inequalities

Quadratic Inequality
An **quadratic inequality** is an inequality that can be written in the form

$$ax^2 + bx + c < 0,$$

for real numbers a, b, and c, with $a \neq 0$. (The symbol $<$ can be replaced with $>$, $\leq$, or $\geq$.)

Solving a Quadratic Inequality
Step 1   Solve the corresponding quadratic equation.

Step 2   Identify the intervals determined by the solutions of the equation.

Step 3   Use a test value from each interval to determine which intervals form the solution set.

EXAMPLE 5 Solving a Quadratic Inequality
Solve $x^2 - x - 12 < 0$.

EXAMPLE 6 Solving a Quadratic Inequality
Solve $2x^2 + 5x - 12 \geq 0$.

Inequalities that use the symbols < and > are _____ inequalities, while ≤ and ≥ are used in _____ inequalities.

EXAMPLE 7 Finding Projectile Height
If a projectile is launched from ground level with an initial velocity of 96 ft per sec, its height s in feet t seconds after launching is given by the following equation.

$$s = -16t^2 + 96t$$

When will the projectile be greater than 80 ft above ground level?

Rational Inequalities

Solving a Rational Inequality

Step 1   Rewrite the inequality, if necessary, so that 0 is on one side and there is a single fraction on the other side.

Step 2   Determine the values that will cause either the numerator or the denominator of the rational expression to equal 0. These values determine the intervals on the number line to consider.

Step 3   Use a test value from each interval to determine which intervals form the solution set.

A value causing a denominator to equal zero will never be included in the solution set. If the inequality is strict, any value causing the numerator to equal zero will be excluded. If the inequality is nonstrict, any such value will be included.

EXAMPLE 8   Solving a Rational Inequality

Solve $\dfrac{5}{x+4} \geq 1$.

Be careful with the endpoints of the intervals when solving rational inequalities.

EXAMPLE 9 Solving a Rational Inequality

Solve $\dfrac{2x-1}{3x+4} < 5$.

Reflect: *When solving a rational inequality, how do you know whether to use an open interval or a closed interval?*

1.8 Absolute Value Equations and Inequalities
■ Basic Concepts ■ Absolute Value Equations ■ Absolute Value Inequalities
■ Special Cases ■ Absolute Value Models for Distance and Tolerance

Key Terms: tolerance

Basic Concepts

Solving Absolute Value Equations and Inequalities

Absolute Value Equation or Inequality	Equivalent Form	Graph of the Solution Set	Solution Set
Case 1: $\|x\| = k$		$-k \quad k$	
Case 2: $\|x\| < k$		$-k \quad k$	
Case 3: $\|x\| > k$		$-k \quad k$	

The equivalent form of $|a| = |b|$ *is* _____ *or* _____.

Absolute Value Equations

EXAMPLE 1 Solving Absolute Value Equations (Case 1 and the Special Case $|a| = |b|$.)

Solve each equation.

(a) $|5 - 3x| = 12$ (b) $|4x - 3| = |x + 6|$

Absolute Value Inequalities

EXAMPLE 2 Solving Absolute Value Inequalities (Cases 2 and 3)
Solve each inequality.
(a) $|2x+1| < 7$ (b) $|2x+1| > 7$

Cases 1, 2, and 3 require that the absolute value expression be isolated on one side of the equation or inequality.

EXAMPLE 3 Solving an Absolute Value Inequality (Case 3)
Solve $|2-7x| - 1 > 4$.

Special Cases

The three cases given in this section require the constant k to be positive. **When $k \leq 0$, use the fact that the absolute value of any expression must be nonnegative, and consider the truth of the statement.**

EXAMPLE 4 Solving Special Cases

Solve each equation or inequality.

(a) $|2 - 5x| \geq -4$

(b) $|4x - 7| < -3$

(c) $|5x + 15| = 0$

Absolute Value Models for Distance and Tolerance

EXAMPLE 5 Using Absolute Value Inequalities to Describe Distances
Write each statement using an absolute value inequality.
(a) k is no less than 5 units from 8.   (b) n is within 0.001 unit of 6.

EXAMPLE 6 Using Absolute Value to Model Tolerance
In quality control and other applications, we often wish to keep the difference between two quantities within some predetermined amount, called the **tolerance**. Suppose $y = 2x + 1$ and we want y to be within 0.01 unit of 4. For what values of x will this be true?

Reflect: Give an example of when tolerance is used.

Chapter 2 Graphs and Functions

2.1 Rectangular Coordinates and Graphs
- Ordered Pairs ■ The Rectangular Coordinate System ■ The Distance Formula
- The Midpoint Formula ■ Graphing Equations in Two Variables

Key Terms: ordered pair, origin, x-axis, y-axis, rectangular (Cartesian) coordinate system, coordinate plane (xy-plane), quadrants, coordinates, collinear, graph of an equation, x-intercept, y-intercept

Ordered Pairs

EXAMPLE 1 Writing Ordered Pairs
Use the table to write ordered pairs to express the relationship between each category and the amount spent on it.

Category	Amount Spent
food	$ 6443
housing	$17,109
transportation	$ 8604
health care	$ 2976
apparel and services	$ 1801
entertainment	$ 2835

Source: U.S. Bureau of Labor Statistics.

(a) housing

(b) entertainment

The Rectangular Coordinate System
The x-axis and y-axis together make up a _____ _____ system, or
_____ _____ system. The plane into which the coordinate system is
introduced is the _____ _____, or _____ _____.
The x-axis and y-axis divide the plane into four regions, or _____.

The Distance Formula

Distance Formula
Suppose that $P(x_1, y_1)$ and $R(x_2, y_2)$ are two points in a coordinate plane. The distance between P and R, written $d(P,R)$, is given by the following formula.

$$d(P,R) = \underline{\hspace{4in}}.$$

The distance between two points in a coordinate plane is the square root of the sum of the square of the difference between their x-coordinates and the square of the difference between their y-coordinates.

EXAMPLE 2 Using the Distance Formula
Find the distance between $P(-8, 4)$ and $Q(3, -2)$.

If the sides a, b, and c of a triangle satisfy $a^2 + b^2 = c^2$ then the triangle is a _____ _____ with _____ having lengths a and b and _____ having length c.

EXAMPLE 3 Applying the Distance Formula
Are the points $M(-2, 5)$, $N(12, 3)$, and $Q(10, -11)$ the vertices of a right triangle?

CHAPTER 2 Graphs and Functions

Reflect: What does the Pythagorean Theorem tell us about right triangles?

EXAMPLE 4 Applying the Distance Formula
Are the points $(-1, 5), (2, -4),$ and $(4, -10)$ collinear?

Reflect: Finish this sentence: Three points are collinear if_____.

The Midpoint Formula

MIDPOINT FORMULA
The coordinates of the midpoint M of the line segment with endpoints $P(x_1, y_1)$ and $Q(x_2, y_2)$ are given by the following.

_____

That is, the x-coordinate of the midpoint of a line segment is the _____ of the x-coordinates of the segment's endpoints, and the y-coordinate is the _____ of the y-coordinates of the segment's endpoints.

EXAMPLE 5 Using the Midpoint Formula
Use the midpoint formula to do each of the following.
(a) Find the coordinates of the midpoint M of the segment with endpoints $(8, -4)$ and $(6, 1)$.

(b) Find the coordinates of the other endpoint Q of a segment with one endpoint $P(-6, 12)$ and midpoint $M(8, -2)$.

EXAMPLE 6 Applying the Midpoint Formula
Figure 8 depicts how a graph might indicate the increase in the revenue generated by fast-food restaurants in the United States from $69.8 billion in 1990 to $164.8 billion in 2010. Use the midpoint formula and the two given points to estimate the revenue from fast-food restaurants in 2000, and compare it to the actual figure of $107.1 billion.

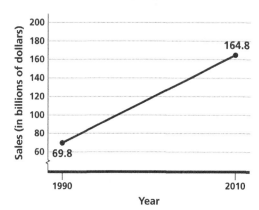

Revenue of Fast-Food Restaurants in U.S.

Source: National Restaurant Association.

Graphing Equations in Two Variables

EXAMPLE 7 Finding Ordered-Pair Solutions of Equations
For each equation, find at least three ordered pairs that are solutions.

(a) $y = 4x - 1$

(b) $x = \sqrt{y - 1}$

(c) $y = x^2 - 4$

Graphing an Equation by Point Plotting

Step 1 Find the intercepts.

Step 2 Find as many additional ordered pairs as needed.

Step 3 Plot the ordered pairs from Steps 1 and 2.

Step 4 Join the points from Step 3 with a smooth line or curve.

EXAMPLE 8 GRAPHING EQUATIONS
Graph each of the equations here, from **Example 7**.

(a) $y = 4x - 1$

(b) $x = \sqrt{y - 1}$

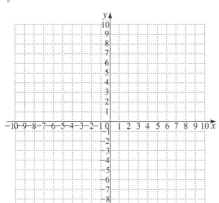
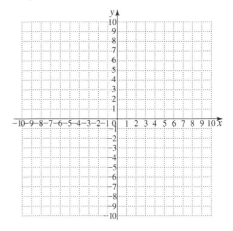

(c) $y = x^2 - 4$

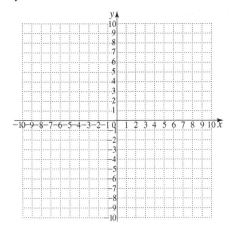

Reflect: *What are intercepts? How do you find the x-intercept(s)? How do you find the y-intercept(s)?*

2.2 Circles
■ Center-Radius Form ■ General Form ■ An Application

Key Terms: circle, radius, center of a circle

Center-Radius Form
By definition, a **circle** is the set of all points in a plane that lie a given distance from a given point. The given distance is called the _____ and the given point is the _____.

Center-Radius Form of the Equation of a Circle
A circle with center (h, k) and radius r has the equation
$$(x - h)^2 + (y - k)^2 = r^2,$$
which is the **center-radius form** of the equation of a circle. A circle with center $(0, 0)$ and radius r has the following equation.
$$x^2 + y^2 = r^2$$

EXAMPLE 1 Finding the Center-Radius Form
Find the center-radius form of the equation of each circle described.
(a) center at $(-3, 4)$, radius 6
(b) center at $(0, 0)$, radius 3

EXAMPLE 2 Graphing Circles
Graph each circle discussed in **Example 1**.

(a) $(x+3)^2 + (y-4)^2 = 36$

(b) $x^2 + y^2 = 9$

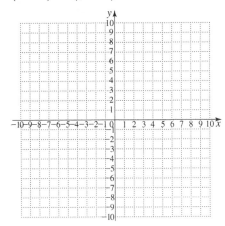
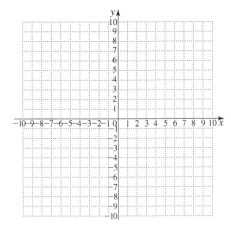

General Form

General Form of the Equation of a Circle
For some real numbers c, d, and e, the equation
$$x^2 + y^2 + cx + dy + e = 0,$$
can have a graph that is a circle or a point, or is nonexistent.

EXAMPLE 3 Finding the Center and Radius by Completing the Square
Show that $x^2 - 6x + y^2 + 10y + 25 = 0$ has a circle as its graph. Find the center and radius.

EXAMPLE 4 Finding the Center and Radius by Completing the Square
Show that $2x^2 + 2y^2 - 6x + 10y = 1$ has a circle as its graph. Find the center and radius.

CHAPTER 2 Graphs and Functions

EXAMPLE 5 Determining Whether a Graph is a Point or Nonexistent

The graph of the equation $x^2 + 10x + y^2 - 4y + 33 = 0$ is either a point or is nonexistent. Which is it?

Reflect: How do you determine whether the graph of an equation is a circle, a point, or is nonexistent?

EXAMPLE 6 Locating the Epicenter of an Earthquake

Suppose receiving stations A, B, and C are located on a coordinate plane at the points $(1, 4)$, $(-3, -1)$, and $(5, 2)$. Let the distances from the earthquake epicenter to these stations be 2 units, 5 units, and 4 units, respectively. Where on the coordinate plane is the epicenter located?

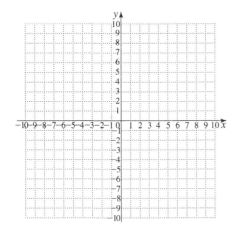

2.3 Functions
- Relations and Functions ■ Domain and Range
- Determining Whether Relations are Functions ■ Function Notation
- Increasing, Decreasing, and Constant Functions

Key Terms: dependent variable, independent variable, relation, function, domain, range, function notation, increasing function, decreasing function, constant function

Relations and Functions
We can often describe one quantity in terms of another:
- The letter grade you receive in a mathematics course depends on your _____.

- The amount you pay for gas at the gas station depends on the number of _____.

- The dollars spent on entertainment depends on the _____.

Relation and Function
A **relation** is a set of _____ _____. A **function** is a relation in which, for each distinct value of the first component of the ordered pairs, there is _____ _____ value of the second component.

EXAMPLE 1 Deciding Whether Relations Define Functions
Decide whether each relation defines a function.
$$F = \{(1,2),(-2,4),(3,4)\}$$
$$G = \{(1,1),(1,2),(1,3),(2,3)\}$$
$$H = \{(-4,1),(-2,1),(-2,0)\}$$

In a function, no two ordered pairs can have the same _____ _____ and different _____ _____.

In a function, there is exactly one value of the _____ variable, the second component, for each value of the _____ variable, the first component.

MN-108   CHAPTER 2 Graphs and Functions

Domain and Range

Domain and Range
In a relation consisting of ordered pairs (x, y), the set of all values of the independent variable (x) is the _____. The set of all values of the dependent variable (y) is the _____.

EXAMPLE 2 Finding Domains and Ranges of Relations
Give the domain and range of each relation. Tell whether the relation defines a function.

(a) $\{(-3,1),(4,2),(4,5),(6,8)\}$

(b)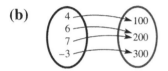

(c)

x	y
-5	2
0	2
5	2

EXAMPLE 3 Finding Domains and Ranges from Graphs
Give the domain and range of each relation.

(a)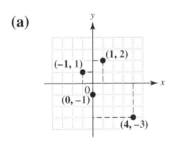

(b)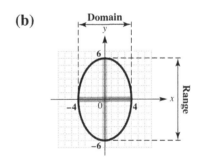

(c)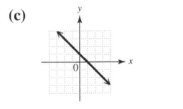

(d)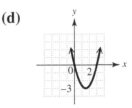

Reflect: *How do you determine the domain of a relation? How do you determine the range?*

Section 2.3 MyNotes   MN-109

Agreement on Domain
Unless specified otherwise, the domain of a relation is assumed to be all real numbers that produce real numbers when substituted for the independent variable.

Determining Whether Relations Are Functions

Vertical Line Test
If every vertical line intersects a graph in no more than _____ _____, then the relation is a function.

EXAMPLE 4 Using the Vertical Line Test
Use the vertical line test to determine whether each relation graphed in **Example 3** is a function.

(a)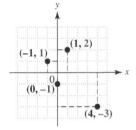

(b)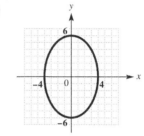

(c)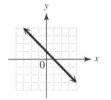

(d)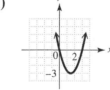

EXAMPLE 5 Identifying Functions, Domains, and Ranges
Decide whether each relation defines a function and give the domain and range.

(a) $y = x + 4$ (b) $y = \sqrt{2x - 1}$ (c) $y^2 = x$

(d) $y \leq x - 1$ (e) $y = \dfrac{5}{x - 1}$

Variations of the Definition of Function
1. A **function** is a relation in which, for each distinct value of the first component of the ordered pairs, there is exactly one value of the second component.
2. A **function** is a set of ordered pairs in which no first component is repeated.
3. A **function** is a rule or correspondence that assigns exactly one range value to each distinct domain value.

Function Notation
When a function f is defined with a rule or an equation using x and y for the independent and dependent variables, we say, "y is a function of x" to emphasize that y *depends on x*. We use the notation

$$y = f(x)$$

called **function notation,** to express this and read $f(x)$ as **"f of x."**

Note that $f(x)$ is just another name for the _____ _____ *y.*

The symbol $f(x)$ does not indicate "f times x," but represents the y-value for the indicated x-value.

EXAMPLE 6 Using Function Notation
Let $f(x) = -x^2 + 5x - 3$ and $g(x) = 2x + 3$. Find and simplify each of the following.

(a) $f(2)$ **(b)** $f(q)$ **(c)** $g(a+1)$

EXAMPLE 7 Using Function Notation

For each function, find $f(3)$.

(a) $f(x) = 3x - 7$

(b) $f = \{(-3, 5), (0, 3)), (3, 1), (6, -1)\}$

(c)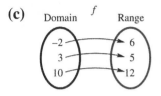

(d)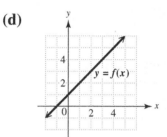

Finding an Expression for $f(x)$

Consider an equation involving x and y. Assume that y can be expressed as a function f of x. To find an expression for $f(x)$, use the following steps.

Step 1 Solve the equation for _____.

Step 2 Replace _____ with _____.

EXAMPLE 8 Writing Equations Using Function Notation

Assume that y is a function f of x. Rewrite each equation using function notation. Then find $f(-2)$ and $f(a)$.

(a) $y = x^2 + 1$

(b) $x - 4y = 5$

Increasing, Decreasing, and Constant Functions

Increasing, Decreasing, and Constant Functions
Suppose that a function f is defined over an *open* interval I and x_1 and x_2 are in I.

(a) f **increases** on I if, whenever $x_1 < x_2$, $f(x_1) < f(x_2)$.

(b) f **decreases** on I if, whenever $x_1 < x_2$, $f(x_1) > f(x_2)$.

(c) f is **constant** on I if, for every x_1 and x_2, $f(x_1) = f(x_2)$.

EXAMPLE 9 Determining Intervals over Which a Function Is Increasing, Decreasing, or Constant

The figure shows the graph of a function. Determine the intervals over which the function is increasing, decreasing, or constant.

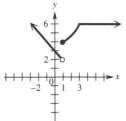

EXAMPLE 10 Interpreting a Graph

Figure 30 (shown to the right) shows the relationship between the number of gallons, $g(t)$, of water in a small swimming pool and time in hours, t. By looking at this graph of the function, we can answer questions about the water level in the pool at various times. For example, at time 0 the pool is empty. The water level then increases, stays constant for a while, decreases, then becomes constant again. Use the graph to respond to the following.

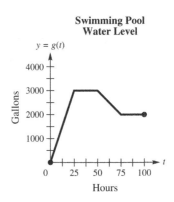

(a) What is the maximum number of gallons of water in the pool? When is the maximum water level first reached?

(b) For how long is the water level increasing? decreasing? constant?

(c) How many gallons of water are in the pool after 90 hr?

(d) Describe a series of events that could account for the water level changes shown in the graph.

MN-114   CHAPTER 2 Graphs and Functions

2.4 Linear Functions
- Graphing Linear Functions ■ Standard Form $Ax + By = C$ ■ Slope
- Average Rate of Change ■ Linear Models

Key Terms: linear function, standard form, relatively prime, change in x, change in y, slope, average rate of change, linear cost function, cost, fixed cost, revenue function, profit function

Graphing Linear Functions

Linear Function
A function f is a **linear function** if

$$f(x) = ax + b,$$

for real numbers a and b. If $a \neq 0$, the domain and range of a linear function are both $(-\infty, \infty)$.

EXAMPLE 1 Graphing a Linear Function Using Intercepts
Graph $f(x) = -2x + 6$. Give the domain and range.

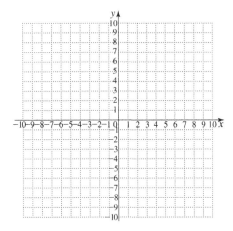

EXAMPLE 2 Graphing a Horizontal Line

Graph $f(x) = -3$. Give the domain and range.

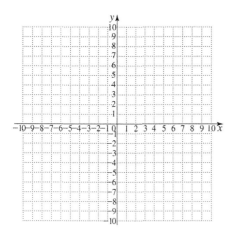

EXAMPLE 3 Graphing a Vertical Line

Graph $x = -3$. Give the domain and range of this relation.

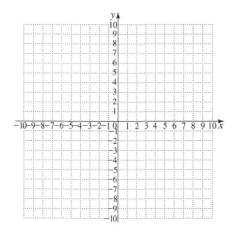

Reflect: *How do you recognize the equation of a horizontal line? How do you recognize the equation of a vertical line?*

MN-116   CHAPTER 2 Graphs and Functions

Standard Form $Ax + By = C$

EXAMPLE 4 Graphing $Ax + By = C$ $(C = 0)$
Graph $4x - 5y = 0$. Give the domain and range.

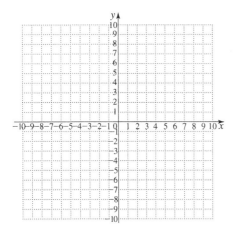

Reflect: How do you recognize the equation of a line that goes through the origin?

Slope

Slope
The **slope** m of the line through the points (x_1, y_1) and (x_2, y_2) is given by the following.

$$m = \frac{\text{rise}}{\text{run}} = \frac{\Delta y}{\Delta x} = \frac{y_2 - y_1}{x_2 - x_1}, \quad \text{where } \Delta x \neq 0$$

That is, the slope of a line is the change in _____ *divided by the corresponding change in* _____*, where the change in* _____ *is not 0.*

When using the slope formula, it makes no difference which point is (x_1, y_1) *or* (x_2, y_2). *However, be consistent. Be sure to write the difference of the* _____ *in the numerator and the difference of the* _____ *in the denominator.*

Reflect: Why can't the slope of a vertical line be found?

Undefined Slope
The slope of a vertical line is _____.

EXAMPLE 5 Finding Slopes With The Slope Formula
Find the slope of the line through the given points.

(a) $(-4, 8)$, $(2, -3)$ (b) $(2, 7)$, $(2, -4)$ (c) $(5, -3)$, $(-2, -3)$

Zero Slope
The slope of a horizontal line is _____.

Slope is the same no matter which pair of distinct points on the line are used to find it.

EXAMPLE 6 Finding the Slope from an Equation
Find the slope of the line $4x + 3y = 12$.

Reflect: *If the equation of a line is in the form $y = ax + b$, the slope is the same as the coefficient of _____ in the equation.*

EXAMPLE 7 Graphing a Line Using a Point and the Slope

Graph the line passing through $(-1, 5)$ and having slope $-\dfrac{5}{3}$.

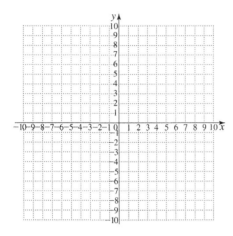

Reflect: *A line with a positive slope rises from _____ to _____.*
A line with a negative slope falls from _____ to _____.
A line with _____ neither rises nor falls.

Average Rate of Change

EXAMPLE 8 Interpreting Slope as Average Rate of Change

In 1980, the average monthly rate for basic cable TV in the United States was $7.69. In 2008, the average monthly rate was $46.13. Assume a linear relationship, and find the average rate of change in the monthly rate per year. Graph as a line segment and interpret the result. (*Source*: SNL Kagan.)

Linear Models

EXAMPLE 9 Writing Linear Cost, Revenue, and Profit Functions

Assume that the cost to produce an item is a linear function and all items produced are sold. The fixed cost is $1500, the variable cost per item is $100, and the item sells for $125. Write linear functions to model

(a) cost, **(b)** revenue, and **(c)** profit

(d) How many items must be sold for the company to make a profit?

2.5 Equations of Lines and Linear Models
- Point-Slope Form ■ Slope-Intercept Form ■ Vertical and Horizontal Lines
- Parallel and Perpendicular Lines ■ Modeling Data
- Solving Linear Equations in One Variable by Graphing

Key Terms: point-slope form, slope-intercept form, scatter diagram

Point-Slope Form

Point-Slope Form
The **point-slope form** of the equation of the line with slope m passing through the point (x_1, y_1) is

$$y - y_1 = m(x - x_1).$$

EXAMPLE 1 Using the Point-Slope Form (Given a Point and the Slope)
Write an equation of the line through $(-4, 1)$ having slope -3.

EXAMPLE 2 Using the Point-Slope Form (Given Two Points)
Write an equation of the line through $(-3, 2)$ and $(2, -4)$. Write the result in standard form $Ax + By = C$.

Slope-Intercept Form

Slope-Intercept Form
The **slope-intercept form** of the equation of the line with slope m and y-intercept b is
$$y = mx + b.$$

EXAMPLE 3 Finding the Slope and y-Intercept from an Equation of a Line
Find the slope and y-intercept of the line with equation $4x + 5y = -10$.

Reflect: The slope m of the graph of $Ax + By = C$ is _____, and the y-intercept is _____.

EXAMPLE 4 Using the Slope-Intercept Form (Given Two Points)
Write an equation of the line through (1, 1) and (2, 4). Then graph the line using the slope-intercept form.

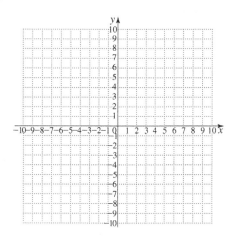

CHAPTER 2 Graphs and Functions

EXAMPLE 5 Finding an Equation from a Graph
Use the graph of the linear function f shown here to complete the following.

(a) Find the slope, y-intercept, and x-intercept.

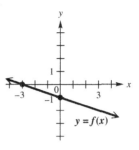

(b) Write the equation that defines f.

Vertical and Horizontal Lines

Equations of Vertical and Horizontal Lines
An equation of the **vertical line** through the point (a,b) is _____.

An equation of the **horizontal line** through the point (a,b) is _____.

Parallel and Perpendicular Lines

Parallel Lines
Two distinct nonvertical lines are parallel if and only if they have the same _____.

Perpendicular Lines
Two lines, neither of which is vertical, are perpendicular if and only if their slopes have a product of _____. Thus, the slopes of perpendicular lines, neither of which is vertical, are _____ _____.

EXAMPLE 6 Finding Equations of Parallel and Perpendicular Lines
Write the equation in both slope-intercept and standard form of the line that passes through the point (3, 5) and satisfies the given condition.

(a) parallel to the line $2x + 5y = 4$

(b) perpendicular to the line $2x + 5y = 4$

Equation	Description	When to Use
$y = mx + b$	**Slope-Intercept Form** Slope is _____. y-intercept is _____.	The slope and y-intercept can be easily identified and used to quickly graph the equation. This form can also be used to find the equation of a line given a point and the slope.
$y - y_1 = m(x - x_1)$	**Point-Slope Form** Slope is _____. Line passes through _____.	This form is ideal for finding the equation of a line if the slope and a point on the line or two points on the line are known.
$Ax + By = C$	**Standard Form** (If the coefficients and constant are rational, then A, B, and C are expressed as relatively prime integers, with $A \geq 0$) Slope is _____. $(B \neq 0)$. x-intercept is _____. $(A \neq 0)$. y-intercept is _____. $(B \neq 0)$.	The x- and y-intercepts can be found quickly and used to graph the equation. The slope must be calculated.
$y = b$	**Horizontal Line** Slope is _____. y-intercept is _____.	If the graph intersects only the y-axis, then y is the only variable in the equation.
$x = a$	**Vertical Line** Slope is _____. x-intercept is _____.	If the graph intersects only the x-axis, then x is the only variable in the equation.

MN-124   CHAPTER 2 Graphs and Functions

Modeling Data

EXAMPLE 7 Finding an Equation of a Line That Models Data

Average annual tuition and fees for in-state students at public four-year colleges are shown in the table for selected years and graphed as ordered pairs of points in Figure 49, where $x = 0$ represents 2005, $x = 1$ represents 2006, and so on, and y represents the cost in dollars. This graph of ordered pairs of data is called a **scatter diagram**.

Year	Cost (in dollars)
2005	5492
2006	5804
2007	6191
2008	6591
2009	7050
2010	7605

Source: Trends in College Pricing 2010, The College Board.

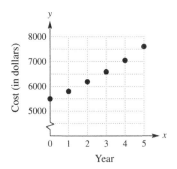

Figure 49

(a) Find an equation that models the data.

(b) Use the equation from part (a) to predict the cost of tuition and fees at public four-year colleges in 2012.

Guidelines for Modeling
Step 1 Make a _____ _____ of the data.
Step 2 Find an _____ that models the data. For a line, this involves selecting _____ data points and finding the _____ of the _____ through them.

Solving Linear Equations in One Variable by Graphing

EXAMPLE 8 **Solving an Equation with a Graphing Calculator**

Use a graphing calculator to solve $-2x - 4(2 - x) = 3x + 4$.

Reflect: Explain how to use a graphing calculator to solve a linear equation.

2.6 Graphs of Basic Functions

- Continuity ■ The Identity, Squaring, and Cubing Function
- The Square Root and Cube Root Functions ■ The Absolute Value Function
- Piecewise-Defined Functions ■ The Relation $x = y^2$

Key Terms: continuous function, parabola, vertex, piecewise-defined function, step function

Continuity

Continuity (Informal Definition)
A function is **continuous** over an interval of its domain if its hand-drawn graph over that interval can be sketched without lifting the _____ from the _____.

If a function is not continuous at a *point,* then it has a *discontinuity* there.

EXAMPLE 1 Determining Intervals of Continuity
Describe the intervals of continuity for each function shown below.

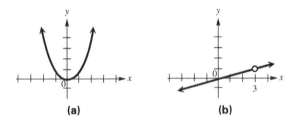

(a)   (b)

The Identity, Squaring, and Cubing Functions

Identity Function $f(x) = x$

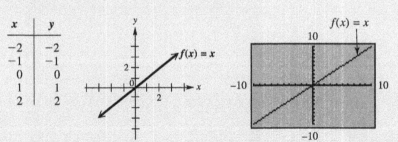

Domain: _____

Range: _____

$f(x) = x$ is increasing on its entire domain, _____.

It is continuous on its entire domain, _____.

Section 2.6 MyNotes   MN-127

Squaring Function $f(x) = x^2$

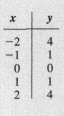 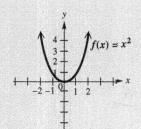 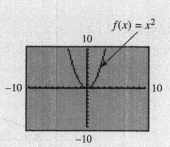

Domain: _____

Range: _____

$f(x) = x^2$ decreases on the interval _____ and increases on the interval _____.

It is continuous on its entire domain, _____.

Cubing Function $f(x) = x^3$

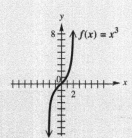 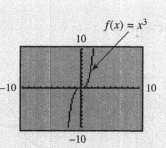

Domain: _____

Range: _____

$f(x) = x^3$ increases on the open interval, _____.

It is continuous on its entire domain, _____.

The Square Root and Cube Root Functions

Square Root Function $f(x) = \sqrt{x}$

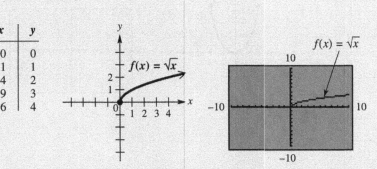

Domain: _____

Range: _____

$f(x) = \sqrt{x}$ increases on the open interval, _____.

It is continuous on its entire domain, _____.

Cube Root Function $f(x) = \sqrt[3]{x}$

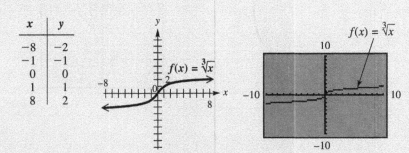

Domain: _____

Range: _____

$f(x) = \sqrt[3]{x}$ increases on its entire domain, _____.

It is continuous on its entire domain, _____.

The Absolute Value Function

Absolute Value Function $f(x) = |x|$

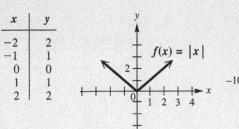
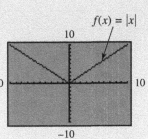

Domain: _____

Range: _____

$f(x) = |x|$ decreases on the open interval _____ and increases on the open interval _____.

It is continuous on its entire domain, _____.

Piecewise-Defined Functions

EXAMPLE 2 Graphing Piecewise-Defined Functions
Graph each function.

(a) $f(x) = \begin{cases} -2x + 5 & \text{if } x \leq 2 \\ x + 1 & \text{if } x > 2 \end{cases}$

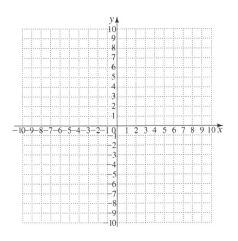

MN-130   CHAPTER 2 Graphs and Functions

(b) $f(x) = \begin{cases} 2x+3 & \text{if } x \leq 0 \\ -x^2+3 & \text{if } x > 0 \end{cases}$

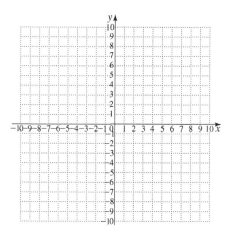

$f(x) = [\![x]\!]$

The **greatest integer function,** $f(x) = [\![x]\!]$, pairs every real number x with the greatest integer _____ _____ or _____ _____ x.

Greatest Integer Function $f(x) = [\![x]\!]$

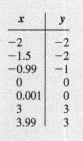 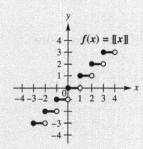 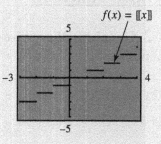

Domain: _____

Range: _____

$f(x) = x$ is constant on the open intervals ..., _____, ...

It is discontinuous at all integer values in its domain, _____.

Section 2.6 MyNotes   MN-131

EXAMPLE 3 Graphing a Greatest Integer Function

Graph $f(x) = \left[\!\left[\dfrac{1}{2}x + 1\right]\!\right]$.

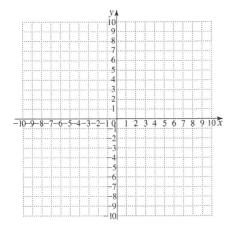

EXAMPLE 4 Applying a Greatest Integer Function

An express mail company charges $25 for a package weighing up to 2 lb. For each additional pound or fraction of a pound there is an additional charge of $3. Let $y = D(x)$ represent the cost to send a package weighing x pounds. Graph $y = D(x)$ for x in the interval $(0, 6]$.

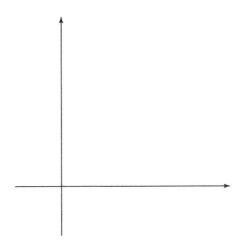

Reflect: *What mathematical symbol is used to identify the greatest integer function? Explain how you find the greatest integer function value of 1.5.*

The Relation $x = y^2$

Recall that a function is a relation where every domain value is paired with _____ _____ _____ _____ range value.

$x = y^2$

Selected Ordered Pairs for $x = y^2$

x	y
0	0
1	± 1
4	± 2
9	± 3

There are two different y-values for the same x-value.

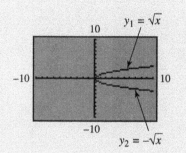

Domain: _____

Range: _____

2.7 Graphing Techniques
■ Stretching and Shrinking ■ Reflecting ■ Symmetry ■ Even and Odd Functions
■ Translations

Key Terms: symmetry, even function, odd function, vertical translation, horizontal translation

Stretching and Shrinking

EXAMPLE 1 Stretching or Shrinking a Graph
Graph each function.

(a) $g(x) = 2|x|$

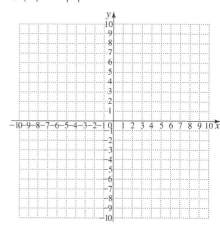

(b) $h(x) = \frac{1}{2}|x|$

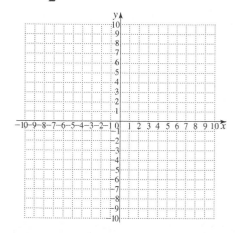

(c) $k(x) = |2x|$

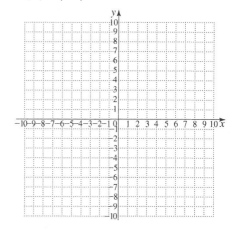

Reflect: How does each graph in Example 1 compare to the graph of the basic function $f(x) = |x|$?

CHAPTER 2 Graphs and Functions

Vertical Stretching or Shrinking of the Graph of a Function
Suppose that $a > 0$. If a point (x, y) lies on the graph of $y = f(x)$, then the point (x, ay) lies on the graph of $y = af(x)$.

(a) If $a > 1$, then the graph of $y = af(x)$ is a _____ _____ of the graph of $y = f(x)$.

(b) If $0 < a < 1$, then the graph of $y = af(x)$ is a _____ _____ of the graph of $y = f(x)$.

Horizontal Stretching or Shrinking of the Graph of a Function
Suppose that $a > 0$. If a point (x, y) lies on the graph of $y = f(x)$, then the point $\left(\frac{x}{a}, y\right)$ lies on the graph of $y = f(ax)$.

(a) If $0 < a < 1$, then the graph of $y = f(ax)$ is a _____ _____ of the graph of $y = f(x)$.

(b) If $a > 1$, then the graph of $y = f(ax)$ is a _____ _____ of the graph of $y = f(x)$.

Reflecting

EXAMPLE 2 Reflecting a Graph Across an Axis
Graph each function.

(a) $g(x) = -\sqrt{x}$

(b) $h(x) = \sqrt{-x}$

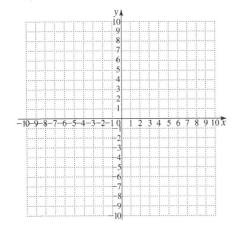
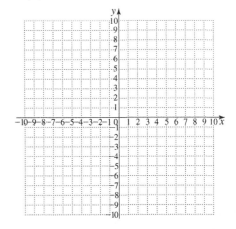

Reflect: *How does each graph in Example 2 compare to the graph of the basic function* $f(x) = \sqrt{x}$ *?*

Reflecting across an Axis
The graph of $y = -f(x)$ is the same as the graph of $y = f(x)$ reflected across the
_____-_____. (If a point (x, y) lies on the graph of $y = f(x)$, then _____ lies on this reflection.)

The graph of $y = f(-x)$ is the same as the graph of $y = f(x)$ reflected across the
_____-_____. (If a point (x, y) lies on the graph of $y = f(x)$, then _____ lies on this reflection.)

Symmetry

Symmetry with Respect to an Axis
The graph of an equation is **symmetric with respect to the** _____-_____ if the replacement of x with $-x$ results in an equivalent equation.

The point _____ is on the graph whenever the point (x, y) is on the graph.

The graph of an equation is **symmetric with respect to the** _____-_____ if the replacement of y with $-y$ results in an equivalent equation.

The point _____ is on the graph whenever the point (x, y) is on the graph.

EXAMPLE 3 Testing for Symmetry with Respect to an Axis
Test for symmetry with respect to the x-axis and the y-axis.

(a) $y = x^2 + 4$ (b) $x = y^2 - 3$

(c) $x^2 + y^2 = 16$ **(d)** $2x + y = 4$

Symmetry with Respect to the Origin
The graph of an equation is **symmetric with respect to the origin** if the replacement of both _____ with _____ and _____ with _____ at the same time results in an equivalent equation.

EXAMPLE 4 Testing for Symmetry with Respect to the Origin
Are the following graphs symmetric with respect to the origin?

(a) $x^2 + y^2 = 16$ **(b)** $y = x^3$

Tests for Symmetry

	Symmetry with Respect to:		
	x-axis	y-axis	Origin
Equation is unchanged if:	_____ is replaced with _____	_____ is replaced with _____	_____ is replaced with _____ and _____ is replaced with _____
Example:			

Even and Odd Functions

Even and Odd Functions
A function f is called an **even function** if _____ for all x in the domain of f. (Its graph is symmetric with respect to the _____.)

A function f is called an **odd function** if _____ for all x in the domain of f. (Its graph is symmetric with respect to the _____.)

EXAMPLE 5 Determining Whether Functions Are Even, Odd, or Neither
Determine whether each function defined is *even, odd,* or *neither*.

(a) $f(x) = 8x^4 - 3x^2$ (b) $f(x) = 6x^3 - 9x$ (c) $f(x) = 3x^2 + 5x$

Translations

EXAMPLE 6 Translating a Graph Vertically
Graph $g(x) = |x| - 4$.

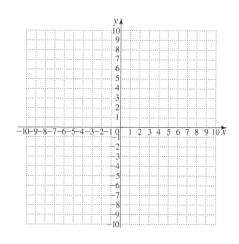

Reflect: *How does the graph in Example 6 compare to the graph of the basic function* $f(x)=|x|$?

Vertical Translations

If a function g is defined by $g(x) = f(x) + c,$ where c is a real number, then for every point (x, y) on the graph of f, there will be a corresponding point _____ on the graph of g.

The graph of g will be the same as the graph of f, but translated c units _____ if c is positive or $|c|$ units _____ if c is negative. The graph of g is called a _____ _____ of the graph of f.

EXAMPLE 7 Translating a Graph Horizontally

Graph $g(x) = |x - 4|$.

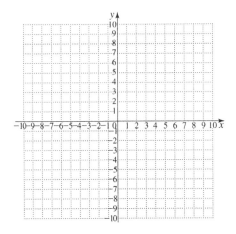

Reflect: *How does the graph in Example 7 compare to the graph of the basic function* $f(x)=|x|$?

Horizontal Translations

If a function g is defined by $g(x) = f(x-c)$, where c is a real number, then for every point (x, y) on the graph of f there will be a corresponding point _____ on the graph of g.

The graph of g will be the same as the graph of f, but translated c units _____ if c is positive or $|c|$ units _____ if c is negative. The graph of g is called a _____ _____ of the graph of f.

(c > 0) To Graph:	Shift the Graph of $y = f(x)$ by c Units:
$y = f(x) + c$	
$y = f(x) - c$	
$y = f(x + c)$	
$y = f(x - c)$	

EXAMPLE 8 Using More Than One Transformation
Graph each function.

(a) $f(x) = -|x+3| + 1$

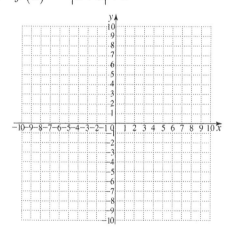

(b) $h(x) = |2x - 4|$

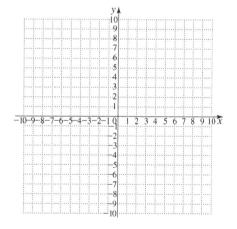

(c) $g(x) = -\dfrac{1}{2}x^2 + 4$

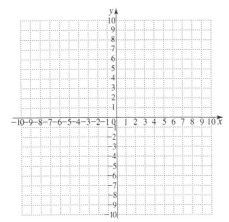

EXAMPLE 9 Graphing Translations of a Given Graph

A graph of a function defined by $y = f(x)$ is shown in the figure. Use this graph to sketch each of the following graphs.

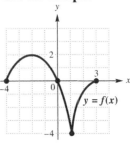

(a) $g(x) = f(x) + 3$

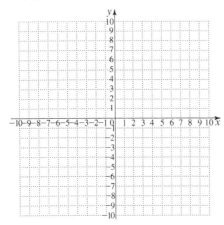

(b) $h(x) = f(x+3)$

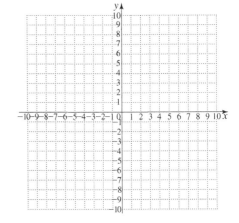

(c) $k(x) = f(x-2)+3$

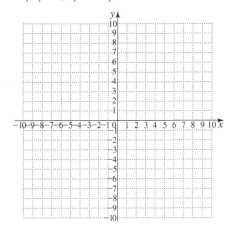

Summary of Graphing Techniques
In the descriptions that follow, assume that $a > 0$, $h > 0$, and $k > 0$. In comparison with the graph of $y = f(x)$:

1. The graph of $y = f(x) + k$ is translated _____ units _____.

2. The graph of $y = f(x) - k$ is translated _____ units _____.

3. The graph of $y = f(x + h)$ is translated _____ units _____.

4. The graph of $y = f(x - h)$ is translated _____ units _____.

5. The graph of $y = af(x)$ is a _____ _____ of the graph of $y = f(x)$ if $a > 1$. It is a _____ _____ if $0 < a < 1$.

6. The graph of $y = f(ax)$ is a _____ _____ of the graph of $y = f(x)$ if $0 < a < 1$. It is a _____ _____ if $a > 1$.

7. The graph of $y = -f(x)$ is _____ across the _____-_____.

8. The graph of $y = f(-x)$ is _____ across the _____-_____.

2.8 Function Operations and Composition
- Arithmetic Operations on Functions ■ The Difference Quotient
- Composition of Functions and Domain

Key Terms: difference quotient, composite function (composition)

Arithmetic Operations on Functions

Operations on Functions and Domains
Given two functions f and g, then for all values of x for which both $f(x)$ and $g(x)$ are defined, the functions $f+g$, $f-g$, fg, and $\frac{f}{g}$ are defined as follows.

$$(f+g)(x) = f(x) + g(x)$$
$$(f-g)(x) = f(x) - g(x)$$
$$(fg)(x) = f(x) \cdot g(x)$$
$$\left(\frac{f}{g}\right)(x) = \frac{f(x)}{g(x)}, \quad g(x) \neq 0$$

The **domains** of $f+g$, $f-g$, and fg include all real numbers in the intersection of the domains of f and g, while the **domain** of $\frac{f}{g}$ includes those real numbers in the intersection of the domains of f and g for which $g(x) \neq 0$.

EXAMPLE 1 Using Operations on Functions
Let $f(x) = x^2 + 1$ and $g(x) = 3x + 5$. Find each of the following.

(a) $(f+g)(1)$

(b) $(f-g)(-3)$

(c) $(fg)(5)$

(d) $\left(\dfrac{f}{g}\right)(0)$

Section 2.8 MyNotes   MN-143

EXAMPLE 2 Using Operations on Functions and Determining Domains
Let $f(x) = 8x - 9$ and $g(x) = \sqrt{2x - 1}$. Find each function in (a)–(d).

(a) $(f + g)(x)$ **(b)** $(f - g)(x)$ **(c)** $(fg)(x)$

(d) $\left(\dfrac{f}{g}\right)(x)$ **(e)** Give the domains of the functions in parts (a) – (d).

EXAMPLE 3 Evaluating Combinations of Functions
If possible, use the given representations of functions f and g to evaluate

$$(f + g)(4), \quad (f - g)(-2), \quad (fg)(1), \quad \text{and} \quad \left(\dfrac{f}{g}\right)(0).$$

(a)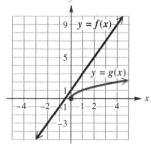

(b)

x	$f(x)$	$g(x)$
-2	-3	undefined
0	1	0
1	3	1
4	9	2

(c) $f(x) = 2x + 1, \ g(x) = \sqrt{x}$

CHAPTER 2 Graphs and Functions

The Difference Quotient

EXAMPLE 4 Finding The Difference Quotient
Let $f(x) = 2x^2 - 3x$. Find and simplify the expression for the difference quotient,
$$\frac{f(x+h) - f(x)}{h}.$$

Composition of Functions and Domain

Composition of Functions and Domain
If f and g are functions, then the **composite function**, or **composition**, of g and f is defined by
$$(g \circ f)(x) = g(f(x)).$$

The **domain of** $(g \circ f)$ is the set of all numbers x in the domain of f such that $f(x)$ is in the domain of g.

EXAMPLE 5 Evaluating Composite Functions
Let $f(x) = 2x - 1$ and $g(x) = \dfrac{4}{x-1}$.

(a) Find $(f \circ g)(2)$
(b) Find $(g \circ f)(-3)$

Copyright © 2015 Pearson Education, Inc.

EXAMPLE 6 Determining Composite Functions and Their Domains

Given that $f(x) = \sqrt{x}$ and $g(x) = 4x + 2$, find each of the following.

(a) $(f \circ g)(x)$ and its domain

(b) $(g \circ f)(x)$ and its domain

Reflect: Explain the process for determining the domain of a composite function.

EXAMPLE 7 Determining Composite Functions and Their Domains

Given that $f(x) = \dfrac{6}{x-3}$ and $g(x) = \dfrac{1}{x}$, find each of the following.

(a) $(f \circ g)(x)$ and its domain

(b) $(g \circ f)(x)$ and its domain

EXAMPLE 8 Showing That $(g \circ f)(x)$ Is Not Equivalent to $(f \circ g)(x)$

Let $f(x) = 4x + 1$ and $g(x) = 2x^2 + 5x$. Show that $(g \circ f)(x) \neq (f \circ g)(x)$. (This is sufficient to prove that this inequality is true in general.)

EXAMPLE 9 Finding Functions That Form a Given Composite

Find functions f and g such that

$$(f \circ g)(x) = (x^2 - 5)^3 - 4(x^2 - 5) + 3.$$

Chapter 3 Polynomial and Rational Functions

3.1 Quadratic Functions and Models
■ Quadratic Functions ■ Graphing Techniques ■ Completing the Square
■ The Vertex Formula ■ Quadratic Models

Key Terms: polynomial function, leading coefficient, leading term, zero polynomial, quadratic function, parabola, axis (axis of symmetry), vertex, quadratic regression

Polynomial Function
A **polynomial function** of degree n, where n is a nonnegative integer, is a function given by

$$f(x) = \underline{\hspace{5cm}}$$

where $a_n, a_{n-1}, \ldots, a_1,$ and a_0 are real numbers with $a_n \neq 0$.

Polynomial Function	Function Name	Degree n	Leading Coefficient a_n
$f(x) = 2$			
$f(x) = 5x - 1$			
$f(x) = 4x^2 - x + 1$			
$f(x) = 2x^3 - \frac{1}{2}x + 5$			
$f(x) = x^4 + \sqrt{2}x^3 - 3x^2$			

Quadratic Functions

Quadratic Function
A function f is a **quadratic function** if

$$f(x) = \underline{\hspace{5cm}},$$

where a, b, and c are real numbers, with $a \neq 0$.

MN-148 CHAPTER 3 Polynomial and Rational Functions

Graphing Techniques

$$F(x) = a(x - h)^2 + k$$

- Opens up if $a > 0$
- Opens down if $a < 0$
- Vertically stretched (narrower) if $|a| > 1$
- Vertically shrunk (wider) if $0 < |a| < 1$

Horizontal shift:
- h units right if $h > 0$
- $|h|$ units left if $h < 0$

Vertical shift:
- k units up if $k > 0$
- $|k|$ units down if $k < 0$

EXAMPLE 1 Graphing Quadratic Functions
Graph each quadratic function. Give the domain and range.

(a) $f(x) = x^2 - 4x - 2$

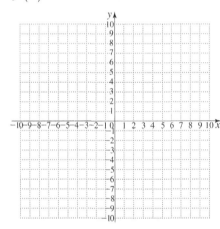

(b) $g(x) = -\dfrac{1}{2}x^2$

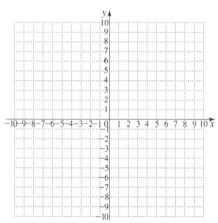

(c) $F(x) = -\dfrac{1}{2}(x - 4)^2 + 3$

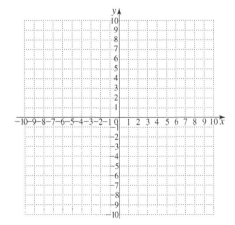

Copyright © 2015 Pearson Education, Inc.

Completing the Square

EXAMPLE 2 Graphing a Parabola by Completing the Square (*a* = 1)

Graph $f(x) = x^2 - 6x + 7$ by completing the square and locating the vertex. Find the largest intervals over which the function is increasing or decreasing.

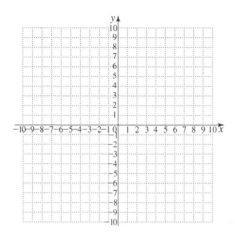

Reflect: How do you determine whether a quadratic function opens up or down?

EXAMPLE 3 Graphing a Parabola by Completing the Square $(a \neq 1)$

Graph $f(x) = -3x^2 - 2x + 1$ by completing the square and locating the vertex. Identify the intercepts of the graph.

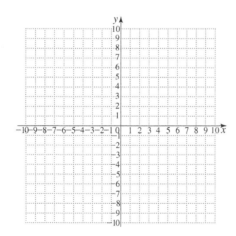

Reflect: How do you determine the x-intercepts of a quadratic function? How do you determine the y-intercept of a quadratic function?

CHAPTER 3 Polynomial and Rational Functions

The Vertex Formula

The vertex form of $f(x) = ax^2 + bx + c$ is given by _____.
It is not necessary to memorize the expression for k, since it is equal to
$f(h) = $ _____ .

Graph of a Quadratic Function
The quadratic function $f(x) = ax^2 + bx + c$ can be written as

$$y = f(x) = \underline{\hspace{2cm}}, \quad a \neq 0,$$

where $h = $ _____ and $k = $ _____ .

The graph of f has the following characteristics.

1. It is a _____ with vertex _____ and the vertical line _____ as axis.
2. It opens _____ if $a > 0$ and _____ if $a < 0$.
3. It is _____ than the graph of $y = x^2$ if $|a| < 1$ and _____ if $|a| > 1$.
4. The y-intercept is _____ .
5. The x-intercepts are found by solving the equation _____ .
   - If $b^2 - 4ac > 0$, the x-intercepts are _____ .
   - If $b^2 - 4ac = 0$, the x-intercept is _____ .
   - If $b^2 - 4ac < 0$, there are _____ _____ .

EXAMPLE 4 Using the Vertex Formula
Find the axis and vertex of the parabola having equation $f(x) = 2x^2 + 4x + 5$.

Quadratic Models

EXAMPLE 5 Solving a Problem Involving Projectile Motion
A ball is thrown directly upward from an initial height of 100 ft with an initial velocity of 80 ft per sec.

(a) Give the function that describes the height of the ball in terms of time t.

(b) After how many seconds does the projectile reach its maximum height? What is this maximum height?

(c) For what interval of time is the height of the ball greater than 160 ft?

(d) After how many seconds will the ball hit the ground?

EXAMPLE 6 Modeling the Number of Hospital Outpatient Visits

The number of hospital outpatient visits (in millions) for selected years is shown in the table. In the table, 95 represents 1995, 100 represents 2000, and so on, and the number of outpatient visits is given in millions.

Year	Visits	Year	Visits
95	483.2	102	640.5
96	505.5	103	648.6
97	520.6	104	662.1
98	545.5	105	673.7
99	573.5	106	690.4
100	592.7	107	693.5
101	612.0	108	710.0

Source: American Hospital Association.

(a) Prepare a scatter diagram, and determine a quadratic model for these data.

(b) Use the model from part (a) to predict the number of visits in 2012.

3.2 Synthetic Division
- **Synthetic Division**
- **Evaluating Polynomial Functions Using the Remainder Theorem**
- **Testing Potential Zeros**

Key Terms: synthetic division, zero of a polynomial function, root (or solution) of an equation

Division Algorithm
Let $f(x)$ and $g(x)$ be polynomials with $g(x)$ of lesser degree than $f(x)$ and $g(x)$ of degree 1 or more. There exist unique polynomials $q(x)$ and $r(x)$ such that

$$f(x) = \underline{\hspace{6cm}}$$

where either $r(x) = 0$ or the degree of $r(x)$ is less than the degree of $g(x)$.

Synthetic Division

*To avoid errors, use **0** as the coefficient for any missing terms, including a missing constant, when setting up the division.*

EXAMPLE 1 Using Synthetic Division
Use synthetic division to divide
$$\frac{5x^3 - 6x^2 - 28x - 2}{x + 2}$$

CHAPTER 3 Polynomial and Rational Functions

Special Case of the Division Algorithm
For any polynomial $f(x)$ and any complex number k, there exists a unique polynomial $q(x)$ and number r such that the following holds.

$$f(x) = (x-k)q(x) + r$$

Evaluating Polynomial Functions Using the Remainder Theorem

Remainder Theorem
If the polynomial $f(x)$ is divided by $x-k$, then the remainder is equal to _____.

EXAMPLE 2 Applying the Remainder Theorem
Let $f(x) = -x^4 + 3x^2 - 4x - 5$. Use the remainder theorem to find $f(-3)$.

Testing Potential Zeros

A **zero** of a polynomial function $f(x)$ is a number k such that $f(k) =$ _____. *The real number zeros are ____-intercepts of the graph of the function.*

EXAMPLE 3 Deciding Whether a Number Is a Zero
Determine whether the given number k is a zero of $f(x)$.
(a) $f(x) = x^3 - 4x^2 + 9x - 6;\ k = 1$

(b) $f(x) = x^4 + x^2 - 3x + 1;\ k = -1$

(c) $f(x) = x^4 - 2x^3 + 4x^2 + 2x - 5;\ k = 1 + 2i$

Reflect: What is the process for determining whether a number is a zero of a polynomial function?

3.3 Zeros of Polynomial Functions
- Factor Theorem ■ Rational Zeros Theorem ■ Number of Zeros
- Conjugate Zeros Theorem ■ Finding Zeros of a Polynomial Function
- Descartes' Rule of Signs

Key Terms: multiplicity of a zero

Factor Theorem

Factor Theorem
For any polynomial function $f(x)$, $x-k$ is a factor of the polynomial if and only if _____.

EXAMPLE 1 Deciding Whether $x - k$ Is a Factor
Determine whether $x-1$ is a factor of each polynomial.

(a) $f(x) = 2x^4 + 3x^2 - 5x + 7$

(b) $f(x) = 3x^5 - 2x^4 + x^3 - 8x^2 + 5x + 1$

EXAMPLE 2 Factoring a Polynomial Given a Zero
Factor $f(x) = 6x^3 + 19x^2 + 2x - 3$ into linear factors if -3 is a zero of f.

Rational Zeros Theorem

Rational Zeros Theorem

If $\dfrac{p}{q}$ is a rational number written in lowest terms, and if $\dfrac{p}{q}$ is a zero of f, a polynomial function with integer coefficients, then p is a factor of the _____ _____, and q is a factor of the _____ _____.

EXAMPLE 3 Using the Rational Zeros Theorem
Consider the polynomial function.

$$f(x) = 6x^4 + 7x^3 - 12x^2 - 3x + 2$$

(a) List all possible rational zeros.

(b) Find all rational zeros and factor $f(x)$ into linear factors.

The rational zeros theorem gives only possible rational zeros. It does not tell us whether these rational numbers are actual zeros.

Reflect: *Explain the relationship between factors of a polynomial function and zeros of a polynomial function.*

MN-158   CHAPTER 3 Polynomial and Rational Functions

Number of Zeros

Fundamental Theorem of Algebra
Every function defined by a polynomial of degree 1 or more has at least one complex zero.

Number of Zeros Theorem
A function defined by a polynomial of degree n has at most ____ distinct zeros.

EXAMPLE 4 Finding a Polynomial Function That Satisfies Given Conditions (Real Zeros)

Find a function f defined by a polynomial of degree 3 that satisfies the given conditions.

(a) Zeros of $-1, 2,$ and 4; $f(1) = 3$

(b) -2 is a zero of multiplicity 3; $f(-1) = 4$

Conjugate Zeros Theorem

Properties of Conjugates
For any complex numbers c and d, the following properties hold.

$$\overline{c+d} = \overline{c}+\overline{d}, \quad \overline{c \cdot d} = \overline{c} \cdot \overline{d}, \quad \text{and} \quad \overline{c^n} = \left(\overline{c}\right)^n$$

Conjugate Zeros Theorem
If $f(x)$ defines a polynomial function *having only real coefficients* and if $z = a+bi$ is a zero of $f(x)$, where a and b are real numbers, then

$$\overline{z} = \underline{\qquad\qquad\qquad} \text{ is also a zero of } f(x).$$

EXAMPLE 5 Finding A Polynomial Function That Satisfies Given Conditions (Complex Zeros)
Find a polynomial function of least degree having only real coefficients and zeros 3 and $2+i$.

Finding Zeros of a Polynomial Function

EXAMPLE 6 Finding All Zeros Given One Zero
Find all zeros of $f(x) = x^4 - 7x^3 + 18x^2 - 22x + 12$, given that $1-i$ is a zero.

Descartes' Rule of Signs

Descartes' Rule of Signs
Let $f(x)$ define a polynomial function with real coefficients and a nonzero constant term, with terms in descending powers of x.

(a) The number of positive real zeros of f either _____ the number of variations in sign occurring in the coefficients of $f(x)$, or is _____ _____ the number of variations by a positive even integer.

(b) The number of negative real zeros of f either _____ the number of variations in sign occurring in the coefficients of $f(-x)$, or is _____ _____ the number of variations by a positive even integer.

EXAMPLE 7 Applying Descartes' Rule of Signs

Determine the different possibilities for the numbers of positive, negative, and nonreal complex zeros of

$$f(x) = x^4 - 6x^3 + 8x^2 + 2x - 1.$$

Descartes' rule of signs does not identify the multiplicity of the zeros of a function.

3.4 Polynomial Functions: Graphs, Applications, and Models

- **Graphs of $f(x) = ax^n$** ■ **Graphs of General Polynomial Functions**
- **Behavior at Zeros** ■ **Turning Points and End Behavior** ■ **Graphing Techniques**
- **Intermediate Value and Boundedness Theorems** ■ **Approximating Real Zeros**
- **Polynomial Models**

Key Terms: turning points, end behavior, dominating term

Graphs of $f(x) = ax^n$

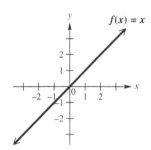
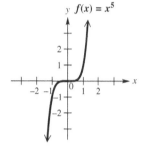

Each figure above has _____ degree and is an _____ function exhibiting symmetry about the _____. Each has domain _____ and range _____ and is continuous on its entire domain _____. Additionally, these _____ functions are _____ on their entire domain _____, appearing as though they _____ to the left and _____ to the right.

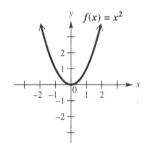
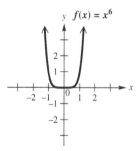

Each figure above has _____ degree and is an _____ function exhibiting symmetry about the _____. Each has domain _____ but restricted range _____. These _____ functions are also continuous on their entire domain _____. However, they are _____ on $(-\infty, 0)$ and _____ on $(0, \infty)$, appearing as though they _____ both to the left and to the right.

Graphs of General Polynomial Functions

Reflect: *Compared to the graph of* $f(x) = ax^n$,

- When $|a| > 1$, the graph is stretched _____, making it _____.

- When $0 < |a| < 1$, the graph is _____ or _____ _____, making it _____.

- The graph of $f(x) = -ax^n$ is _____ across the _____-_____.

- The graph of $f(x) = ax^n + k$ is translated (shifted) _____ units _____ if $k > 0$ and _____ units _____ if $k < 0$.

- The graph of $f(x) = a(x-h)^n$ is translated _____ units _____ if $h > 0$ and _____ units to the _____ if $h < 0$.

- The graph of $f(x) = a(x-h)^n + k$ shows a combination of these _____.

EXAMPLE 1 Examining Vertical and Horizontal Translations

Graph each function. Determine the intervals of the domain for which each function is increasing or decreasing.

(a) $f(x) = x^5 - 2$

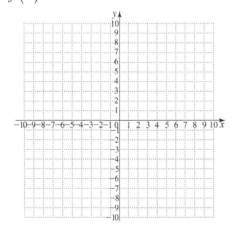

(b) $f(x) = (x+1)^6$

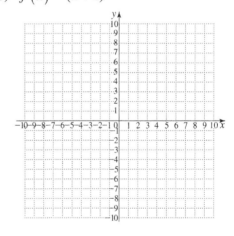

(c) $f(x) = -2(x-1)^3 + 3$

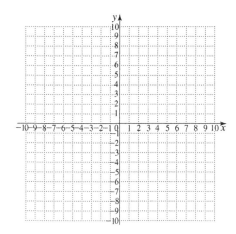

The domain of a polynomial function is the set of _____ _____.
Polynomial functions are smooth, continuous curves on the interval _____. The range of a polynomial of odd degree is also the set of _____ _____.

A polynomial function of even degree has a range of the form _____ or _____, for some real number k.

Behavior at Zeros

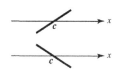

The graph crosses the x-axis at $(c, 0)$ if c is a zero of multiplicity 1.	The graph is tangent to the x-axis at $(c, 0)$ if c is a zero of even multiplicity. The graph bounces, or turns, at c.	The graph crosses **and** is tangent to the x-axis at $(c, 0)$ if c is a zero of odd multiplicity greater than 1. The graph wiggles at c.
(a)	(b)	(c)

Turning Points and End Behavior

Turning Points
A polynomial function of degree n has at most _____ turning points, with at least one turning point between each pair of successive zeros.

MN-164   CHAPTER 3 Polynomial and Rational Functions

End Behavior of Graphs of Polynomial Functions
Suppose that ax^n is the dominating term of a polynomial function f of **odd degree**.
1. If $a > 0$, then as $x \to \infty$, _____, and as $x \to -\infty$, _____.
   We symbolize it as _____.

2. If $a < 0$, then as $x \to \infty$, _____, and as $x \to -\infty$, _____.
   We symbolize it as _____.

Suppose that ax^n is the dominating term of a polynomial function f of **even degree**.
1. If $a > 0$, then as $|x| \to \infty$, _____. We symbolize it as _____.

2. If $a < 0$, then as $|x| \to \infty$, _____. We symbolize it as _____.

EXAMPLE 2 Determining End Behavior
The graphs of the functions defined as follows are shown in A–D.

$$f(x) = x^4 - x^2 + 5x - 4 \qquad g(x) = -x^6 + x^2 - 3x - 4$$
$$h(x) = 3x^3 - x^2 + 2x - 4 \quad \text{and} \quad k(x) = -x^7 + x - 4$$

Based on the discussion of end behavior, match each function in with its graph.

A.
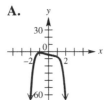

B.
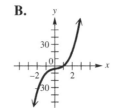

C.
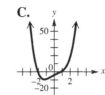

D.
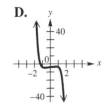

Reflect: *Why does the highest degree term of a polynomial function determine the end behavior of the graph of the polynomial function?*

Graphing Techniques

Graphing a Polynomial Function

Let $f(x) = a_n x^n + a_{n-1} x^{n-1} + \cdots + a_1 x + a_0$, with $a_n \neq 0$, be a polynomial function of degree n. To sketch its graph, follow these steps.

Step 1 _____

Step 2 _____

Step 3 _____

EXAMPLE 3 Graphing a Polynomial Function

Graph $f(x) = 2x^3 + 5x^2 - x - 6$.

Step 1

Step 2

MN-166 CHAPTER 3 Polynomial and Rational Functions

Step 3

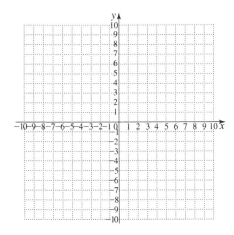

EXAMPLE 4 Graphing a Polynomial Function

Graph $f(x) = -(x-1)(x-3)(x+2)^2$.

Step 1

Step 2

Step 3

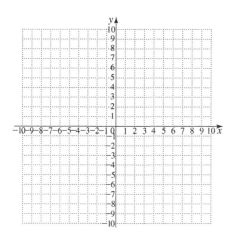

x-Intercepts, Zeros, Solutions, and Factors
If f is a polynomial function and (a, 0) is an x-intercept of the graph of $y = f(x)$, then

a is a _____ of f, a is a _____ of $f(x) = 0$,

and $x - a$ is a _____ of $f(x)$.

Intermediate Value and Boundedness Theorems

Intermediate Value Theorem for Polynomials
If $f(x)$ defines a polynomial function with only real coefficients, and if for real numbers a and b, the values of $f(a)$ and $f(b)$ are opposite in sign, then there exists at least one real _____ between a and b.

EXAMPLE 5 Locating a Zero
Use synthetic division and a graph to show that $f(x) = x^3 - 2x^2 - x + 1$ has a real zero between 2 and 3.

Boundedness Theorem

Let $f(x)$ be a polynomial function of degree $n \geq 1$ with real coefficients and with a positive leading coefficient. Suppose $f(x)$ is divided synthetically by $x - c$.

(a) If $c > 0$ and all numbers in the bottom row of the synthetic division are nonnegative, then $f(x)$ has no zero _____ than c.

(b) If $c < 0$ and the numbers in the bottom row of the synthetic division alternate in sign (with 0 considered positive or negative, as needed), then $f(x)$ has no zero _____ than c.

EXAMPLE 6 Using the Boundedness Theorem

Show that the real zeros of $f(x) = 2x^4 - 5x^3 + 3x + 1$ satisfy these conditions.

(a) No real zero is greater than 3.   (b) No real zero is less than -1.

Approximating Real Zeros

EXAMPLE 7 Approximating Real Zeros of a Polynomial Function

Approximate the real zeros of $f(x) = x^4 - 6x^3 + 8x^2 + 2x - 1$.

Polynomial Models

EXAMPLE 8 Examining a Polynomial Model

The table shows the number of transactions, in millions, by users of bank debit cards for selected years.

Year	Transactions (in millions)
1995	829
1998	3765
2000	5290
2004	14,106
2008	28,464
2011*	39,049

Source: Statistical Abstract of the United States.
*Projected

(a) Using $x = 0$ to represent 1995, $x = 3$ to represent 1998, and so on, use the regression feature of a calculator to determine the quadratic function that best fits the data. Plot the data and the graph.

(b) Repeat part (a) for a cubic function (degree 3).

(c) Repeat part (a) for a quartic function (degree 4).

(d) The **correlation coefficient**, R, is a measure of the strength of the relationship between two variables. The values of R and R^2 are used to determine how well a regression model fits a set of data. The closer the value of R^2 is to 1, the better the fit. Compare R^2 for the three functions found in parts (a)–(c) to decide which function best fits the data.

3.5 Rational Functions: Graphs, Applications, and Models

■ The Reciprocal Function $f(x) = \frac{1}{x}$ ■ The Function $f(x) = \frac{1}{x^2}$ ■ Asymptotes
■ Steps for Graphing Rational Functions ■ Rational Function Models

Key Terms: rational function, discontinuous graph, vertical asymptote, horizontal asymptote, oblique asymptote, point of discontinuity

Rational Function
A function f of the form

$$f(x) = \frac{p(x)}{q(x)},$$

where $p(x)$ and $q(x)$ are polynomials, with $q(x) \neq$ _____, is a **rational function**.

The Reciprocal Function $f(x) = \dfrac{1}{x}$

The Reciprocal Function $f(x) = \dfrac{1}{x}$

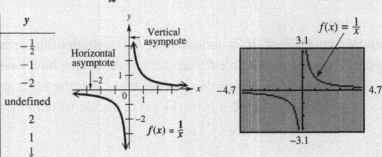

x	y
-2	$-\frac{1}{2}$
-1	-1
$-\frac{1}{2}$	-2
0	undefined
$\frac{1}{2}$	2
1	1
2	$\frac{1}{2}$

Domain: _____

Range: _____

$f(x) = \dfrac{1}{x}$ decreases on the intervals _____ and _____.

It is discontinuous at $x =$ _____.

The ____-axis is a vertical asymptote and the ____-axis is a horizontal asymptote.

It is an _____ function, and its graph is symmetric with respect to the _____.

Section 3.5 MyNotes   MN-171

EXAMPLE 1 Graphing a Rational Function

Graph $y = -\dfrac{2}{x}$. Give the domain and range and the largest intervals of the domain for which the function is increasing or decreasing.

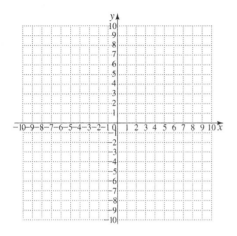

EXAMPLE 2 Graphing a Rational Function

Graph $f(x) = \dfrac{2}{x+1}$. Give the domain and range and the largest intervals of the domain for which the function is increasing or decreasing.

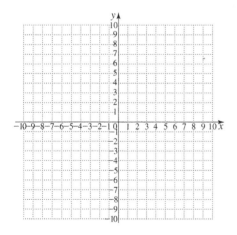

Copyright © 2015 Pearson Education, Inc.

The Function $f(x) = \dfrac{1}{x^2}$

The Function $f(x) = \dfrac{1}{x^2}$

x	y
±3	$\frac{1}{9}$
±2	$\frac{1}{4}$
±1	1
±$\frac{1}{2}$	4
±$\frac{1}{4}$	16
0	undefined

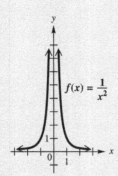
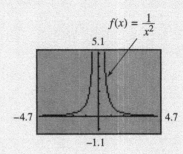

Domain: _____

Range: _____

$f(x) = \dfrac{1}{x^2}$ increases on the interval _____ and decreases on the interval _____.

It is discontinuous at $x =$ ____.

The ___-axis is a vertical asymptote and the ___-axis is a horizontal asymptote.

It is an _____ function, and its graph is symmetric with respect to the _____.

EXAMPLE 3 Graphing a Rational Function

Graph $y = \dfrac{1}{(x+2)^2} - 1$. Give the domain and range and the largest intervals of the domain for which the function is increasing or decreasing.

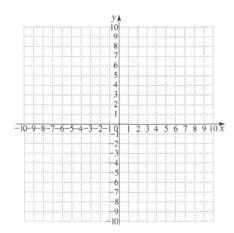

Asymptotes

Asymptotes

Let $p(x)$ and $q(x)$ define polynomials. Consider the rational function $f(x) = \frac{p(x)}{q(x)}$, written in lowest terms, and real numbers a and b.

1. If $|f(x)| \to \infty$ as $x \to a$, then the line $x = a$ is a _____ **asymptote**.
2. If $f(x) \to b$ as $|x| \to \infty$, then the line $y = b$ is a _____ **asymptote**.

Determining Asymptotes
To find the asymptotes of a rational function defined by a rational expression in *lowest terms*, use the following procedures.

1. **Vertical Asymptotes**
   Find any vertical asymptotes by setting the denominator equal to ___ and solving for x. If a is a zero of the denominator, then the **line** _____ **is a vertical asymptote.**

2. **Other Asymptotes**
   Determine other asymptotes by considering these three possibilities:

   (a) If the numerator has lesser degree than the denominator, then there is a **horizontal asymptote** _____ (the x-axis).

   (b) If the numerator and denominator have the same degree, and the function is of the form
   $$f(x) = \frac{a_n x^n + \cdots + a_0}{b_n x^n + \cdots + b_0}, \quad \text{where} \quad a_n, b_n \neq 0,$$
   then the **horizontal asymptote has equation** _____.

   (c) If the numerator is of degree exactly one more than the denominator, then there will be an **oblique (slanted) asymptote**. To find it, divide the numerator by the denominator and disregard the remainder. Set the rest of the quotient equal to y to obtain the equation of the asymptote.

The graph of a rational function may have more than one vertical asymptote, or it may have none at all. *The graph cannot intersect any vertical asymptote. There can be at most one other (nonvertical) asymptote, and the graph may intersect that asymptote.*

EXAMPLE 4 Finding Asymptotes of Rational Functions

For each rational function f, find all asymptotes.

(a) $f(x) = \dfrac{x+1}{(2x-1)(x+3)}$

(b) $f(x) = \dfrac{2x+1}{x-3}$

(c) $f(x) = \dfrac{x^2+1}{x-2}$

Steps for Graphing Rational Functions

GRAPHING A RATIONAL FUNCTION

Let $f(x) = \dfrac{p(x)}{q(x)}$ define a function where $p(x)$ and $q(x)$ are polynomials and the rational expression is written in lowest terms. To sketch its graph, follow these steps.

Step 1   Find any _____ asymptotes.

Step 2   Find any _____ or _____ asymptotes.

Step 3   Find the *y*-intercept by evaluating _____.

Step 4   Find the *x*-intercepts, if any, by solving $f(x) = $ _____. (These will correspond to the zeros of the _____, $p(x)$.)

Step 5   Determine whether the graph will intersect its nonvertical asymptote $y = b$ or $y = mx + b$ by solving $f(x) = $ _____ or $f(x) = $ _____.

Step 6   Plot selected points, as necessary. Choose an *x*-value in each domain interval determined by the vertical asymptotes and *x*-intercepts.

Step 7   Complete the sketch.

EXAMPLE 5 Graphing a Rational Function with the *x*-Axis as Horizontal Asymptote

Graph $f(x) = \dfrac{x+1}{2x^2 + 5x - 3}$.

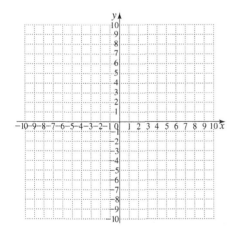

EXAMPLE 6 Graphing a Rational Function That Does Not Intersect Its Horizontal Asymptote

Graph $f(x) = \dfrac{2x+1}{x-3}$.

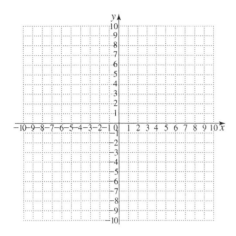

EXAMPLE 7 Graphing a Rational Function That Intersects Its Horizontal Asymptote

Graph $f(x) = \dfrac{3x^2 - 3x - 6}{x^2 + 8x + 16}$.

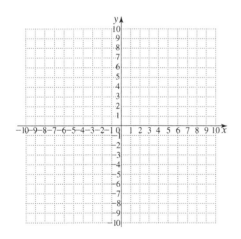

EXAMPLE 8 Graphing A Rational Function with an Oblique Asymptote

Graph $f(x) = \dfrac{x^2+1}{x-2}$

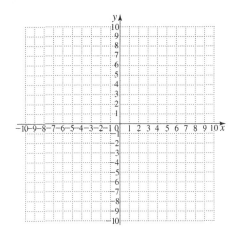

Reflect: *How do you determine the vertical asymptote(s) of the graph of a rational function?*

Reflect: *How do you determine the horizontal or oblique asymptote of a rational function?*

EXAMPLE 9 Graphing A Rational Function Defined by an Expression That Is Not in Lowest Terms

Graph $f(x) = \dfrac{x^2 - 4}{x - 2}$.

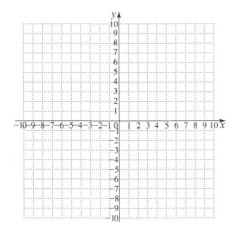

Reflect: What causes the graph of a rational function to have a "hole" in it?

Rational Function Models

EXAMPLE 10 Modeling Traffic Intensity with a Rational Function

Vehicles arrive randomly at a parking ramp at an average rate of 2.6 vehicles per minute. The parking attendant can admit 3.2 vehicles per minute. However, since arrivals are random, lines form at various times. (*Source*: Mannering, F. and W. Kilareski, *Principles of Highway Engineering and Traffic Analysis*, 2nd ed., John Wiley & Sons.)

(a) The **traffic intensity** x is defined as the ratio of the average arrival rate to the average admittance rate. Determine x for this parking ramp.

(b) The average number of vehicles waiting in line to enter the ramp is given by

$$f(x) = \frac{x^2}{2(1-x)},$$

where $0 \leq x < 1$ is the traffic intensity. Graph $f(x)$ and compute $f(0.8125)$ for this parking ramp.

(c) What happens to the number of vehicles waiting as the traffic intensity approaches 1?

3.6 Variation
■ Direct Variation ■ Inverse Variation ■ Combined and Joint Variation

Key Terms: varies directly (directly proportional to), constant of variation, varies inversely (inversely proportional to), combined variation, varies jointly

Direct Variation

Direct Variation
y varies directly as *x*, or *y* is _____ _____ to *x*, if there exists a nonzero real number *k*, called the _____ _____ _____, such that for all *x*,
$$y = kx.$$

Solving Variation Problems
Step 1 Write the general relationship among the variables as an _____. Use the constant _____.

Step 2 Substitute given values of the variables and find the value of *k*.

Step 3 Substitute this value of *k* into the equation from Step 1, obtaining a specific formula.

Step 4 Substitute the remaining values and solve for the required unknown.

EXAMPLE 1 Solving a Direct Variation Problem
The area of a rectangle varies directly as its length. If the area is 50 m^2 when the length is 10 m, find the area when the length is 25 m.

Direct Variation as *n*th Power
Let *n* be a positive real number. Then *y* **varies directly as the *n*th power** of *x*, or *y* is _____ of *x*, if for all *x* there exists a nonzero real number *k* such that
$$y = kx^n.$$

Inverse Variation

Inverse Variation as *n*th Power
Let *n* be a positive real number. Then **y varies inversely as the *n*th power** of *x*, or *y* is
_____ of *x*, if for all *x* there
exists a nonzero real number *k* such that

$$y = \frac{k}{x^n}.$$

If $n = 1$, then $y = \frac{k}{x}$, and *y* _____ _____ as *x*.

EXAMPLE 2 Solving an Inverse Variation Problem
In a certain manufacturing process, the cost of producing a single item varies inversely as the square of the number of items produced. If 100 items are produced, each costs $2. Find the cost per item if 400 items are produced.

Combined and Joint Variation

Joint Variation
Let *m* and *n* be real numbers. Then **y varies jointly** as the *n*th power of *x* and the *m*th power of *z* if for all *x* and *z*, there exists a nonzero real number *k* such that

$$y = kx^n z^m.$$

Note that *and* in the expression "*y* varies jointly as *x* and *z*" translates as the product y = kxz. **The word "and" does not indicate addition here.**

EXAMPLE 3 Solving a Joint Variation Problem

The area of a triangle varies jointly as the lengths of the base and the height. A triangle with base 10 ft and height 4 ft has area 20 ft². Find the area of a triangle with base 3 ft and height 8 ft.

EXAMPLE 4 Solving a Combined Variation Problem

The number of vibrations per second (the pitch) of a steel guitar string varies directly as the square root of the tension and inversely as the length of the string. If the number of vibrations per second is 50 when the tension is 225 newtons and the length is 0.60 m, find the number of vibrations per second when the tension is 196 newtons and the length is 0.65 m.

Reflect: *How can you tell if a problem should be solved using a direct variation, an inverse variation, a joint variation, or a combined variation problem?*

Chapter 4 Inverse, Exponential, and Logarithmic Functions

4.1 Inverse Functions
- One-to-One Functions ■ Inverse Functions ■ Equations of Inverses
- An Application of Inverse Functions to Cryptography

Key Terms: one-to-one function, inverse function

One-to-One Functions

In a one-to-one function, each x-value corresponds to only _____ y-value, and each y-value corresponds to only _____ x-value.

One-to-One Function
A function f is a **one-to-one function** if, for elements a and b in the domain of f,

$$a \neq b \quad \text{implies} \quad f(a) \neq f(b)$$

EXAMPLE 1 Deciding Whether Functions Are One-to-One
Decide whether each function is one-to-one.

(a) $f(x) = -4x + 12$

(b) $f(x) = \sqrt{25 - x^2}$

Horizontal Line Test
A function is one-to-one if every _____ _____ intersects the graph of the function at most once.

MN-184 CHAPTER 4 Inverse, Exponential, and Logarithmic Functions

EXAMPLE 2 Using the Horizontal Line Test
Determine whether each graph is the graph of a one-to-one function.

(a) (b)

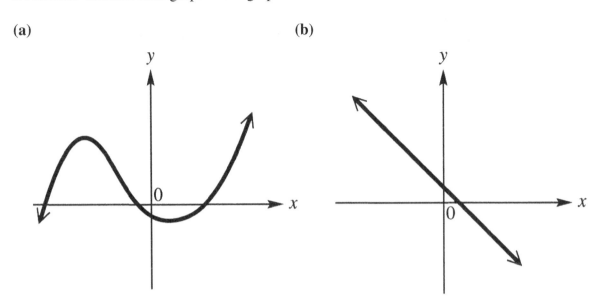

In general, a function that is either _____ *or* _____ *on its entire domain, such as,* $f(x) = -x$, $g(x) = x^3$, *and* $h(x) = \sqrt{x}$ *must be* _____.

Reflect: *How do we determine whether a function is one-to-one?*

Copyright © 2015 Pearson Education, Inc.

Inverse Functions

Inverse Function
Let f be a one-to-one function. Then g is the **inverse function** of f if

$$(f \circ g)(x) = x \qquad \text{for every } x \text{ in the domain of } g,$$

and $\qquad (g \circ f)(x) = x \qquad \text{for every } x \text{ in the domain of } f.$

The condition that f is one-to-one in the definition of inverse function is essential. Otherwise, g will not define a _____.

EXAMPLE 3 Deciding Whether Two Functions are Inverses

Let functions f and g be defined by $f(x) = x^3 - 1$ and $g(x) = \sqrt[3]{x+1}$. Is g the inverse function of f?

By the definition of inverse function, the _____ of f is the _____ of f^{-1}, and the _____ of f is the _____ of f^{-1}.

EXAMPLE 4 Finding Inverses of One-to-One Functions

Find the inverse of each function that is one-to-one.

(a) $F = \{(-2,1),(-1,0),(0,1),(1,2),(2,2)\}$

(b) $G = \{(3,1),(0,2),(2,3),(4,0)\}$

(c) The table in the margin shows the number of days in Illinois that were unhealthy for sensitive groups for selected years using the Air Quality Index (AQI). Let f be the function defined in the table, with the years forming the domain and the numbers of unhealthy days forming the range.

Equations of Inverses

The inverse of a one-to-one function is found by interchanging the x- and y-values of each of its ordered pairs. The equation of the inverse function defined by $y = f(x)$ is found in the same way.

Finding the Equation of the Inverse of $y = f(x)$

For a one-to-one function f defined by an equation $y = f(x)$, find the defining equation of the inverse as follows. (If necessary replace $f(x)$ with y first. Any restrictions on x and y should be considered.)

Step 1 _____

Step 2 _____

Step 3 _____

EXAMPLE 5 Finding Equations of Inverses

Decide whether each equation defines a one-to-one function. If so, find the equation of the inverse.

(a) $f(x) = 2x + 5$

(b) $y = x^2 + 2$

(c) $f(x) = (x-2)^3$

MN-188   CHAPTER 4 Inverse, Exponential, and Logarithmic Functions

EXAMPLE 6 Finding the Equation of the Inverse of a Rational Function

The rational function $f(x) = \dfrac{2x+3}{x-4}$, $x \neq 4$ (Section 3.5) is a one-to-one function. Find its inverse.

EXAMPLE 7 Graphing f^{-1} Given the Graph of f

In each set of axes in **Figure 9**, the graph of a one-to-one function f is shown. Graph f^{-1}. (Graph both f and f^{-1} for each.)

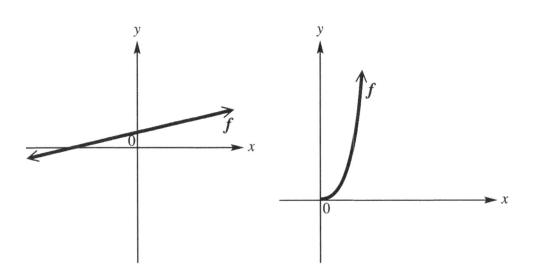

EXAMPLE 8 Finding the Inverse of a Function with a Restricted Domain
Let $f(x) = \sqrt{x+5}$, $x \geq -5$. Find $f^{-1}(x)$.

Important Facts about Inverses
1. If f is one-to-one, then f^{-1} _____.
2. The domain of f is the _____ of f^{-1}, and the range of f is the _____ of f^{-1}.
3. If the point (a,b) lies on the graph of f, then _____ lies on the graph of f^{-1}. The graphs of f and f^{-1} are reflections of each other across the line _____.
4. To find the equation for f^{-1}, replace $f(x)$ with y, interchange __and__, and solve for y. This gives $f^{-1}(x)$.

An Application of Inverse Functions to Cryptography

EXAMPLE 9 Using Functions to Encode and Decode a Message

Use the one-to-one function $f(x) = 3x + 1$ and the following numerical values assigned to each letter of the alphabet to encode and decode the message BE MY FACEBOOK FRIEND.

A	1	H	8	O	15	V	22
B	2	I	9	P	16	W	23
C	3	J	10	Q	17	X	24
D	4	K	11	R	18	Y	25
E	5	L	12	S	19	Z	26
F	6	M	13	T	20		
G	7	N	14	U	21		

4.2 Exponential Functions
■ Exponents and Properties ■ Exponential Functions ■ Exponential Equations
■ Compound Interest ■ The Number *e* and Continuous Compounding
■ Exponential Models

Key Terms: exponential function, exponential equation, compound interest, future value, present value, compound amount, continuous compounding

Exponents and Properties

In this section, we extend the definition of a^r to include all real (not just rational) values of the exponent r.

EXAMPLE 1 Evaluating an Exponential Expression
If $f(x) = 2^x$, find each of the following.

(a) $f(-1)$ (b) $f(3)$ (c) $f\left(\dfrac{5}{2}\right)$ (d) $f(4.92)$

Exponential Functions

Exponential Function
If $a > 0$ and $a \neq 1$, then
$$f(x) = a^x$$
defines the **exponential function with base *a*.**

We do not allow _____ as the base for an exponential function.

Exponential Function $f(x) = a^x$

Domain: _____ Range: _____

For $f(x) = 2^x$:

x	$f(x)$
-2	$\frac{1}{4}$
-1	$\frac{1}{2}$
0	1
1	2
2	4
3	8

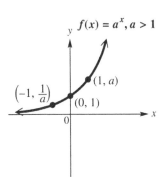

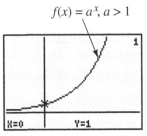

This is the general behavior seen on a calculator graph for any base a, for $a > 1$.

Figure 14

- $f(x) = a^x$, for $a > 1$, is increasing and continuous on its entire domain, _____ .
- The x-axis is a vertical asymptote as $x \to$ _____ .
- The graph passes through the points $\left(-1, \frac{1}{a}\right)$, $(0, \_\_)$, and $(1, \_\_)$.

For $f(x) = \left(\frac{1}{2}\right)^x$:

x	$f(x)$
-3	8
-2	4
-1	2
0	1
1	$\frac{1}{2}$
2	$\frac{1}{4}$

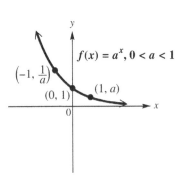

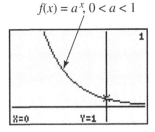

This is the general behavior seen on a calculator graph for any base a, for $0 < a < 1$.

Figure 15

- $f(x) = a^x$, for $0 < a < 1$, is increasing and continuous on its entire domain, _____ .
- The x-axis is a vertical asymptote as $x \to$ _____ .
- The graph passes through the points $\left(-1, \frac{1}{a}\right)$, $(0, \_\_)$, and $(1, \_\_)$.

EXAMPLE 2 Graphing an Exponential Function

Graph $f(x) = 5^x$. Give the domain and range.

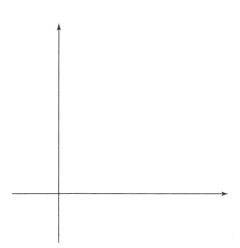

EXAMPLE 3 Graphing Reflections and Translations

Graph each function. Show the graph of $y = 2^x$ for comparison. Give the domain and range.

(a) $f(x) = -2^x$ **(b)** $f(x) = 2^{x+3}$ **(c)** $f(x) = 2^{x-2} - 1$

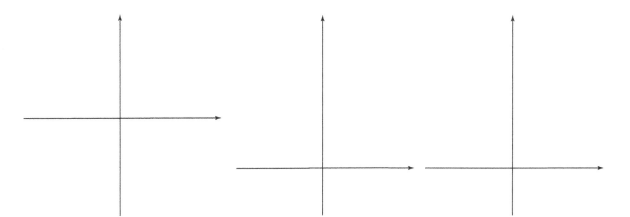

Reflect: *Translations and reflections of exponential functions are the same as for other basic functions. What causes an exponential function to be reflected across the x-axis? Translated horizontally? Translated vertically?*

Exponential Equations

EXAMPLE 4 Solving an Exponential Equation

Solve $\left(\dfrac{1}{3}\right)^x = 81$.

EXAMPLE 5 Solving an Exponential Equation

Solve $2^{x+4} = 8^{x-6}$.

Reflect: *Describe a process for solving exponential equations when the variable is in the exponent as in Example 4 and Example 5.*

EXAMPLE 6 Solving an Equation with a Fractional Exponent
Solve $x^{4/3} = 81$.

Compound Interest
The formula for **compound interest** (interest paid on both principal and interest) is an important application of exponential functions.

Compound Interest
If P dollars are deposited in an account paying an annual rate of interest r compounded (paid) n times per year, then after t years the account will contain A dollars, according to the following formula.

$$A = \left(1 + \frac{r}{n}\right)^{tn}$$

EXAMPLE 7 Using the Compound Interest Formula
Suppose $1000 is deposited in an account paying 4% interest per year compounded quarterly (four times per year).

(a) Find the amount in the account after 10 yr with no withdrawals.

(b) How much interest is earned over the 10-yr period?

EXAMPLE 8   Finding Present Value
Becky Anderson must pay a lump sum of $6000 in 5 yr.

(a) What amount deposited today (present value) at 3.1% compounded annually will grow to $6000 in 5 yr?

(b) If only $5000 is available to deposit now, what annual interest rate is necessary for the money to increase to $6000 in 5 yr?

The Number *e* and Continuous Compounding

The more often interest is compounded within a given time period, the more interest will be earned. Surprisingly, however, there is a limit on the amount of interest, no matter how often it is compounded.

Suppose that $1 is invested at 100% interest per year, compounded *n* times per year. Thus, $A = P\left(1+\frac{r}{n}\right)^{nt}$ becomes $A = \left(1+\frac{1}{n}\right)^n$. What happens to this expression as *n* increases?

n	$A = \left(1+\frac{1}{n}\right)^n$ (rounded)
1	
2	
5	
10	
100	
1000	
10,000	
1,000,000	

Value of *e*
$e \approx$ _____

Continuous Compounding
If *P* dollars are deposited at a rate of interest *r* compounded continuously for *t* years, the compound amount *A* in dollars on deposit is given by the following formula.

$$A = Pe^{rt}$$

CHAPTER 4 Inverse, Exponential, and Logarithmic Functions

EXAMPLE 9 Solving a Continuous Compounding Problem
Suppose $5000 is deposited in an account paying 3% interest compounded continuously for 5 yr. Find the total amount on deposit at the end of 5 yr.

EXAMPLE 10 Comparing Interest Earned as Compounding is More Frequent
In **Example 7**, we found that $1000 invested at 4% compounded quarterly for 10 yr grew to $1488.86. Compare this same investment compounded annually, semiannually, monthly, daily, and continuously.

Compounded	$1	$1000
Annually		
Semiannually		
Quarterly		
Monthly		
Daily		
Continuously		

Exponential Models

EXAMPLE 11 Using Data to Model Exponential Growth

Data from recent past years indicates that future amounts of carbon dioxide in the atmosphere may grow according to the table. Amounts are given in parts per million.

Year	Carbon Dioxide (ppm)
1990	353
2000	375
2075	590
2175	1090
2275	2000

(a) Make a scatter diagram of the data. Do the carbon dioxide levels appear to grow exponentially?

(b) One model for the data is the function $y = 0.001942 e^{0.00609 x}$, where x is the year and $1990 \leq x \leq 2275$. Use a graph of this model to estimate when future levels of carbon dioxide will double and triple over the preindustrial level of 280 ppm.

CHAPTER 4 Inverse, Exponential, and Logarithmic Functions

4.3 Logarithmic Functions
■ Logarithms ■ Logarithmic Equations ■ Logarithmic Functions
■ Properties of Logarithms

Key Terms: logarithm, base, argument, logarithmic equation, logarithmic function

Logarithms

The horizontal line test shows that exponential functions are one-to-one and thus have inverse functions. The equation defining the inverse of a function is found by interchanging x and y in the equation that defines the function. Starting with $y = a^x$ and interchanging x and y yields $x = a^y$.

LOGARITHM
For all real numbers y and all positive numbers a and x, where $a \neq 1$,

$$y = \log_a x \quad \text{is equivalent to} \quad x = a^y$$

The expression $\log_a x$ represents the _____ to which the base a must be raised in order to obtain x.

EXAMPLE 1 Writing Equivalent Logarithmic and Exponential Forms
The table shows several pairs of equivalent statements, written in both logarithmic and exponential forms. Complete the table.

Logarithmic Form	Exponential Form
$\log_2 8 = 3$	
$\log_{1/2} 16 = -4$	
$\log_{10} 100,000 = 5$	
	$3^{-4} = \dfrac{1}{81}$
	$5^1 = 5$
	$\left(\dfrac{3}{4}\right)^0 = 1$

Logarithmic Equations

EXAMPLE 2 Solving Logarithmic Equations
Solve each equation.

(a) $\log_x \dfrac{8}{27} = 3$

(b) $\log_4 x = \dfrac{5}{2}$

(c) $\log_{49} \sqrt[3]{7} = x$

Reflect: Describe a process for solving logarithmic equations as shown in EXAMPLE 2.

Logarithmic Functions

Logarithmic Function
If $a > 0$, $a \neq 1$, and $x > 0$, then

$$f(x) = \log_a x$$

defines the **logarithmic function with base** a.

Logarithmic Function $f(x) = \log_a x$

Domain: _____   Range: _____

For $f(x) = \log_2 x$:

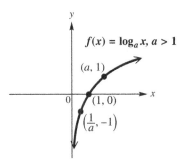
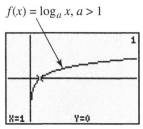

This is the general behavior seen on a calculator graph for **any base a, for $a > 1$.**

Figure 26

- $f(x) = \log_a x$, **for $a > 1$,** is increasing and continuous on its entire domain, _____ .
- The y-axis is a vertical asymptote as $x \to$ ___ from the right.
- The graph passes through the points $(\frac{1}{a}, -1)$, $(1, \_)$, and $(a, \_)$.

For $f(x) = \log_{1/2} x$:

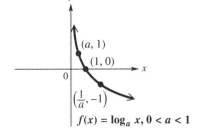
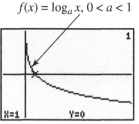

This is the general behavior seen on a calculator graph for **any base a, for $0 < a < 1$.**

Figure 27

- $f(x) = \log_a x$, **for $0 < a < 1$,** is decreasing and continuous on its entire domain, _____ .
- The y-axis is a vertical asymptote as $x \to$ ___ from the right.
- The graph passes through the points $(\frac{1}{a}, -1)$, $(1, \_)$, and $(a, \_)$.

Characteristics of the Graph of $f(x) = \log_a x$

1. The points _____, _____, and _____ are on the graph.
2. If _____, then f is an increasing function. If _____, then f is a decreasing function.
3. The _____ is a vertical asymptote.
4. The domain is _____, and the range is _____.

EXAMPLE 3 Graphing Logarithmic Functions
Graph each function.

(a) $f(x) = \log_{1/2} x$

(b) $f(x) = \log_3 x$

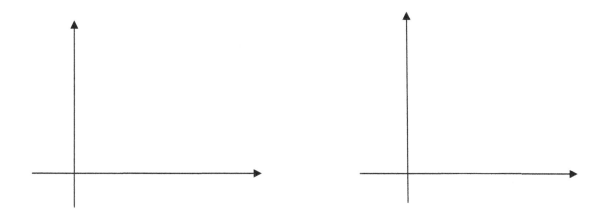

EXAMPLE 4 Graphing Translated Logarithmic functions
Graph each function. Give the domain and range.

(a) $f(x) = \log_2(x-1)$

(b) $f(x) = (\log_3 x) - 1$

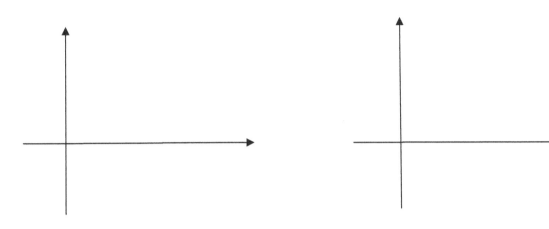

(c) $f(x) = \log_4(x+2) + 1$

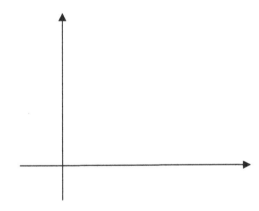

Reflect: *Translations and reflections of logarithmic functions are the same as for other basic functions. What causes a logarithmic function to be reflected across the x-axis? Shifted horizontally? Shifted vertically?*

Properties of Logarithms

The properties of logarithms enable us to change the form of logarithmic statements so that products can be converted into sums, quotients can be converted to differences, and powers can be converted to products.

PROPERTIES OF LOGARITHMS
For $x > 0, y > 0, a > 0, a \neq 1$, and any real number r, the following properties hold.

Property	Description
Product Property $\log_a xy = \log_a x + \log_a y$	The logarithm of the product of two numbers is equal to the _____ of the logarithms of the numbers.
Quotient Property $\log_a \dfrac{x}{y} = \log_a x - \log_a y$	The logarithm of the quotient of two numbers is equal to the _____ between the logarithms of the numbers.
Power Property $\log_a x^r = r \log_a x$	The logarithm of a number raised to a power is equal to the exponent _____ the logarithm of the number.
Logarithm of 1 $\log_a 1 = 0$	The base a logarithm of 1 is _____.
Base a Logarithm of a $\log_a a = 1$	The base a logarithm of a is _____.

EXAMPLE 5 Using the Properties of Logarithms

Rewrite each expression. Assume all variables represent positive real numbers, with $a \neq 1$ and $b \neq 1$.

(a) $\log_6(7 \cdot 9)$

(b) $\log_9 \dfrac{15}{7}$

(c) $\log_5 \sqrt{8}$

(d) $\log_a \dfrac{mnq}{p^2 t^4}$

(e) $\log_a \sqrt[3]{m^2}$

(f) $\log_b \sqrt[n]{\dfrac{x^3 y^5}{z^m}}$

EXAMPLE 6 Using the Properties of Logarithms
Write each expression as a single logarithm with coefficient 1. Assume all variables represent positive real numbers, with $a \neq 1$ and $b \neq 1$.

(a) $\log_3(x+2) + \log_3 x - \log_3 2$

(b) $2\log_a m - 3\log_a n$

(c) $\dfrac{1}{2}\log_b m + \dfrac{3}{2}\log_b 2n - \log_b m^2 n$

There is no property of logarithms to rewrite a logarithm of a sum or difference.

EXAMPLE 7 Using the Properties of Logarithms with Numerical Values
Assume that $\log_{10} 2 = 0.3010$. Find each logarithm.

(a) $\log_{10} 4$

(b) $\log_{10} 5$

THEOREM ON INVERSES
For $a > 0$, $a \neq 1$, the following properties hold.
$$a^{\log_a x} = x \text{ (for } x > 0\text{)} \quad \text{and} \quad \log_a a^x = x$$

4.4 Evaluating Logarithms and the Change-of-Base Theorem
■ Common Logarithms ■ Applications and Models with Common Logarithms ■
■ Natural Logarithms ■ Applications and Models with Natural Logarithms
■ Logarithms with Other Bases

Key Terms: common logarithm, pH, natural logarithm

Common Logarithms

Two of the most most important bases for logarithms are 10 and e. Base 10 logarithms are called **common logarithms.** The common logarithm of x is written $\log x$, where the base is understood to be _____.

Common Logarithm
For all positive numbers x,
$$\log x = \log_{10} x.$$

EXAMPLE 1 Evaluating Common Logarithms with a Calculator
Use a calculator to find the values of

$$\log 1000, \quad \log 142, \quad \text{and} \quad \log 0.005832.$$

Base a logarithms of numbers between _____ and _____, where a > 1, are always negative, as suggested by the graphs in Section 4.3

Applications and Models with Common Logarithms

In chemistry the **pH** of a solution is defined as

$$\text{pH} = -\log\left[H_3O^+\right]$$

where $\left[H_3O^+\right]$ is the hydronium ion concentration in moles per liter. The pH value is a measure of the acidity or alkalinity of a solution. Pure water has a pH 7.0, substances with pH values greater than 7.0 are alkaline, and substances with pH values less than 7.0 are acidic. For ease, we will round pH values to the nearest tenth.

EXAMPLE 2 Finding pH

(a) Find the pH of a solution with $[H_3O^+] = 2.5 \times 10^{-4}$.

(b) Find the hydronium ion concentration of a solution with pH = 7.1.

EXAMPLE 3 Using pH in an Application

Wetlands are classified as bogs, fens, marshes, and swamps based on pH values. A pH value between 6.0 and 7.5 indicates that the wetland is a "rich fen." When the pH is between 3.0 and 6.0, it is a "poor fen," and if the pH falls to 3.0 or less, the wetland is a "bog." (*Source:* R. Mohlenbrock, "Summerby Swamp, Michigan," *Natural History*.)

Suppose that the hydronium ion concentration of a sample of water from a wetland is 6.3×10^{-5}. How would this wetland be classified?

EXAMPLE 4 Measuring the Loudness of Sound

The loudness of sounds is measured in **decibels.** We first assign an intensity of I_0 to a very faint **threshold sound.** If a particular sound has intensity I, then the decibel rating d of this louder sound is given by the following formula.

$$d = 10 \log \frac{I}{I_0}$$

Find the decibel rating of a sound with intensity $10{,}000\, I_0$.

Natural Logarithms

In most practical applications of logarithms, e is used as base. Logarithms with base e are called **natural logarithms,** since they occur in the life sciences and economics in natural situations that involve growth and decay. The base e logarithm of x is written $\ln x$ (read "el-en x"). *The expression $\ln x$ represents the exponent to which e must be raised in order to obtain x.*

Natural Logarithm
For all positive numbers x,
$$\ln x = \log_e x .$$

EXAMPLE 5 Evaluating Natural Logarithms with a Calculator
Use a calculator to find the values of
$$\ln e^3, \quad \ln 142, \quad \text{and} \quad \ln 0.005832.$$

Applications and Models with Natural Logarithms

EXAMPLE 6 Measuring the Age of Rocks
Geologists sometimes measure the age of rocks by using "atomic clocks." By measuring the amounts of potassium-40 and argon-40 in a rock, it is possible to find the age t of the specimen in years with the formula

$$t = (1.26 \times 10^9) \frac{\ln\left(1 + 8.33\left(\frac{A}{K}\right)\right)}{\ln 2},$$

where A and K are the numbers of atoms of argon-40 and potassium-40, respectively, in the specimen.

(a) How old is a rock in which $A = 0$ and $K > 0$?

(b) The ratio $\frac{A}{K}$ for a sample of granite from new Hampshire is 0.212. How old is the sample?

EXAMPLE 7 Modeling Global Temperature Increase

Carbon dioxide in the atmosphere traps heat from the sun. The additional solar radiation trapped by carbon dioxide is called **radiative forcing.** It is measured in watts per square meter (w/m^2). In 1896 the Swedish scientist Svante Arrhenius modeled radiative forcing R caused by additional atmospheric carbon dioxide, using the logarithmic equation

$$R = k \ln \frac{C}{C_0},$$

where C_0 is the preindustrial amount of carbon dioxide, C is the current carbon dioxide level, and k is a constant. Arrhenius determined that $10 \leq k \leq 16$ when $C = 2C_0$. (*Source:* Clime, W., *The Economics of Global Warming,* Institute for International Economics, Washington, D.C.)

(a) Let $C = 2C_0$. Is the relationship between R and k linear or logarithmic?

(b) The average global temperature increase T (in °F) is given by $T(R) = 1.03R$. Write T as a function of k.

Logarithms with Other Bases

We can use a calculator to find the values of either natural logarithms (base e) or common logarithms (base 10). However, sometimes we must use logarithms with other bases. The change-of-base theorem can be used to convert logarithms from one base to another.

Change-of-Base Theorem
For any positive real numbers x, a, and b, where $a \neq 1$ and $b \neq 1$, the following holds.
$$\log_a x = \frac{\log_b x}{\log_b a}$$

EXAMPLE 8 Using the Change-of-Base Theorem
Use the change-of-base theorem to find an approximation to four decimal places for each logarithm.

(a) $\log_5 17$ (b) $\log_2 0.1$

EXAMPLE 9 Modeling Diversity of Species
One measure of the diversity of the species in an ecological community is modeled by the formula

$$H = -[P_1 \log_2 P_1 + P_2 \log_2 P_2 + \cdots + P_n \log_2 P_n],$$

where $P_1, P_2, ..., P_n$ are the proportions of a sample that belong to each of n species found in the sample. (*Source:* Ludwig, J., and J. Reynolds, *Statistical Ecology: A Primer on Methods and Computing,* New York, Wiley.)

Find the measure of diversity in a community with two species where there are 90 of one species and 10 of the other.

4.5 Exponential and Logarithmic Equations
■ Exponential Equations ■ Logarithmic Equations ■ Applications and Models

Exponential Equations

Property of Logarithms
If $x > 0, y > 0, a > 0$, and $a \neq 1$, then the following holds.
$\quad x = y$ is equivalent to $\log_a x = \log_a y$

EXAMPLE 1 Solving an Exponential Equation
Solve $7^x = 12$. Give the solution to the nearest thousandth.

EXAMPLE 2 Solving an Exponential Equation
Solve $3^{2x-1} = 0.4^{x+2}$. Give the solution to the nearest thousandth.

Reflect: When solving exponential equations as in Examples 1 and 2 above, we are able to get the variable out of the exponent by taking what step?

CHAPTER 4 Inverse, Exponential, and Logarithmic Functions

EXAMPLE 3 Solving Base e Exponential Equations
Solve each equation. Give solutions to the nearest thousandth.

(a) $e^{x^2} = 200$

(b) $e^{2x+1} \cdot e^{-4x} = 3e$

EXAMPLE 4 Solving an Exponential Equation Quadratic in Form
Solve $e^{2x} - 4e^x + 3 = 0$. Give exact value(s) for x.

Logarithmic Equations

EXAMPLE 5 Solving Logarithmic Equations
Solve each equation. Give exact values.

(a) $7 \ln x = 28$

(b) $\log_2(x^3 - 19) = 3$

EXAMPLE 6 Solving a Logarithmic Equation
Solve $\log(x+6) - \log(x+2) = \log x$. Give exact value(s).

Recall that the domain of $y = \log_a x$ **is** _____. **For this reason,** *it is always necessary to check that proposed solutions of a logarithmic equation result in logarithms of* _____ *numbers in the original equation.*

EXAMPLE 7 Solving a Logarithmic Equation
Solve $\log_2\left[(3x-7)(x-4)\right]=3$. Give exact value(s).

EXAMPLE 8 Solving a Logarithmic Equation
Solve $\log(3x+2)+\log(x-1)=1$. Give exact value(s).

EXAMPLE 9 Solving a Base e Logarithmic Equation
Solve $\ln e^{\ln x} - \ln(x-3) = \ln 2$. Give exact value(s).

Solving Exponential or Logarithmic Equations

To solve an exponential or logarithmic equation, change the given equation into one of the following forms, where a and b are real numbers, $a > 0$ and $a \neq 1$, and follow the guidelines.

1. $a^{f(x)} = b$
   Solve by _____.
2. $\log_a f(x) = b$
   Solve by _____.
3. $\log_a f(x) = \log_a g(x)$
   The given equation is equivalent to the equation _____. Solve algebraically.
4. In a more complicated equation, such as
$$e^{2x+1} \cdot e^{-4x} = 3e$$
   in **Example 3(b)**, it may be necessary to first solve for $a^{f(x)}$ or $\log_a f(x)$ and then solve the resulting equation using one of the methods given above.
5. Check that the proposed solution is in the domain.

Applications and Models

EXAMPLE 10 Applying an Exponential Equation to the Strength of a Habit

The strength of a habit is a function of the number of times the habit is repeated. If N is the number of repetitions and H is the strength of the habit, then, according to psychologist C. L. Hull,

$$H = 1000\left(1 - e^{-kN}\right),$$

where k is a constant. Solve this equation for k.

EXAMPLE 11 Modeling Coal Consumpiton in the U.S.

The table gives U.S. coal consumption (in quadrillions of British thermal units, or *quads*) for several years. The data can be modeled with the function defined by

$$f(t) = 24.92 \ln t - 93.31, \quad t \geq 80,$$

where t is the number of years after 1900, and $f(t)$ is in quads.

Year	Coal Consumption (in quads)
1980	15.42
1985	17.48
1990	19.17
1995	20.09
2000	22.58
2005	22.80
2008	22.39

Source: U.S. Energy Information Administration.

(a) Approximately what amount of coal was consumed in the United States in 2003? How does this figure compare to the actual figure of 22.32 quads?

(b) If this trend continues, approximately when will annual consumption reach 25 quads?

4.6 Applications and Models of Exponential Growth and Decay
■ The Exponential Growth or Decay Function ■ Growth Function Models
■ Decay Function Models

Key Terms: doubling time, half-life

The Exponential Growth or Decay Function

In many situations that occur in ecology, biology, economics, and the social sciences, a quantity changes at a rate proportional to the amount present. In such cases, the amount present at time t is a special function of t called an **exponential growth or decay function.**

Exponential Growth or Decay Function
Let y_0 be the amount or number present at time $t = 0$. Then, under certain conditions, the amount present at any time t is modeled by
$$y = y_0 e^{kt}, \text{ where } k \text{ is a constant.}$$

Growth Function Models

EXAMPLE 1 Determining an Exponential Function to Model the Increase of Carbon Dioxide
In **Example 11, Section 4.2,** we discussed the growth of atmospheric carbon dioxide over time. A function based on the data from the table was given in that example. Now we can see how to determine such a function from the data.

Year	Carbon Dioxide (ppm)
1990	353
2000	375
2075	590
2175	1090
2275	2000

Source: International Panel on Climate Change (IPCC)

(a) Find an exponential function that gives the amount of carbon dioxide y in year x.

(b) Estimate the year when future levels of carbon dioxide will be double the preindustrial level of 280 ppm.

EXAMPLE 2 Finding Doubling Time for Money
How long will it take for the money in an account that accrues interest at a rate of 3%, compounded continuously, to double?

EXAMPLE 3 Determining an Exponential Function to Model Population Growth

According to the U.S. Census Bureau, the world population reached 6 billion people during 1999 and was growing exponentially. By the end of 2010, the population had grown to 6.947 billion. The projected world population (in billions of people) t years after 2010, is given by the function defined by $f(t) = 6.947e^{0.00745t}$.

(a) Based on this model, what will the world population be in 2020?

(b) In what year will the world population reach 9 billion?

Decay Function Models

EXAMPLE 4 Determining an Exponential Function to Model Radioactive Decay

Suppose 600 g of a radioactive substance are present initially and 3 yr later only 300 g remain,

(a) Determine the exponential equation that models this decay.

(b) How much of the substance will be present after 6 yr?

EXAMPLE 5   Solving a Carbon Dating Problem
Carbon-14, also known as radiocarbon, is a radioactive form of carbon that is found in all living plants and animals. After a plant or animal dies, the radiocarbon disintegrates. Scientists can determine the age of the remains by comparing the amount of radiocarbon with the amount present in living plants and animals. This technique is called **carbon dating.** The amount of radiocarbon present after t years is given by $y = y_0 e^{-0.0001216t}$, where y_0 is the amount present in living plants and animals.

(a) Find the half-life of carbon-14.

(b) Charcoal from an ancient fire pit on Java contained ¼ the carbon-14 of a living sample of the same size. Estimate the age of the charcoal.

Section 4.6 MyNotes   MN-225

EXAMPLE 6 Modeling Newton's Law of Cooling

Newton's law of cooling says that the rate at which a body cools is proportional to the difference in temperature between the body and the environment around it. The temperature $f(t)$ of the body at time t in appropriate units after being introduced into an environment having constant temperature T_0 is

$$f(t) = T_0 + Ce^{-kt}, \quad \text{where } C \text{ and } k \text{ are constants.}$$

A pot of coffee with a temperature of 100°C is set down in a room with a temperature of 20°C. The coffee cools to 60°C after 1 hr.

(a) Write an equation to model the data.

(b) Find the temperature after half an hour.

(c) How long will it take for the coffee to cool down to 50°C?

Copyright © 2015 Pearson Education, Inc.

Chapter 5 Systems and Matrices

5.1 Systems of Linear Equations
■ Linear Systems ■ Substitution Method ■ Elimination Method ■ Special Systems
■ Applying Systems of Equations
■ Solving Linear Systems with Three Unknowns (Variables)
■ Using Systems of Equations to Model Data

Key Terms: linear equation (first-degree equation) in n unknowns, system of equations, solutions of a system of equations, system of linear equations (linear system), consistent system, independent equations, inconsistent system, dependent equations, equivalent systems, ordered triple

Linear Systems

3 Cases for Solutions of Systems

1. **The graphs intersect at exactly one point,** which gives the (single) ordered-pair solution of the system. The **system is** _____ and the **equations are** _____.

2. **The graphs are parallel lines,** so there is no solution and the solution set is _____. The **system is** _____ and the **equations are** _____.

3. **The graphs are the same line,** and there are an infinite number of solutions. The **system is** _____ and the **equations are** _____.

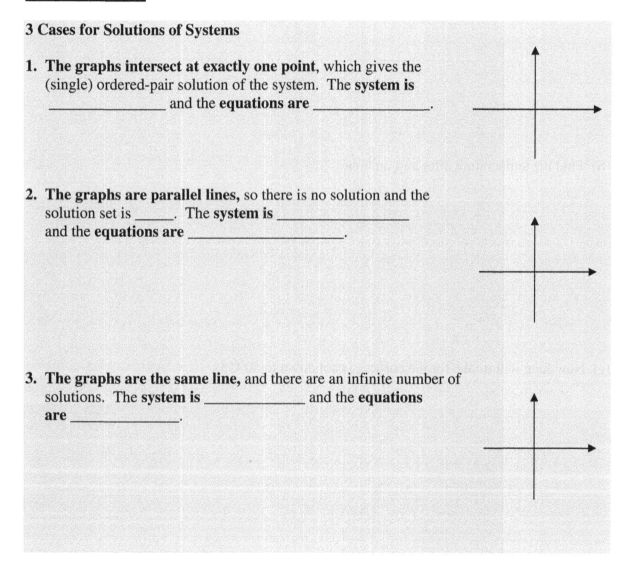

Substitution Method

EXAMPLE 1 Solving a System by Substitution
Solve the system.
$$3x + 2y = 11$$
$$-x + y = 3$$

Elimination Method

Transformations of a Linear System

1. Interchange any two equations of the system.

2. Multiply or divide any equation of the system by a nonzero real number.

3. Replace any equation of the system by the sum of that equation and a multiple of another equation in the system.

EXAMPLE 2 Solving a System by Elimination
Solve the system.
$$3x - 4y = 1$$
$$2x + 3y = 12$$

Special Systems

EXAMPLE 3 Solving an Inconsistent System
Solve the system.
$$3x - 2y = 4$$
$$-6x + 4y = 7$$

Reflect: *How do you recognize an inconsistent linear system?*

EXAMPLE 4 Solving a System with Infinitely Many Solutions

Solve the system.

$$8x - 2y = -4$$
$$-4x + y = 2$$

Reflect: *How do you recognize a linear system with infinitely many solutions?*

Applying Systems of Equations

Solving an Applied Problem by Writing a System of Equations

Step 1 _____ the problem carefully until you understand what is given and what is to be found.

Step 2 _____ to represent the unknown values, using diagrams or tables as needed. *Write down* what each variable represents.

Step 3 _____ that relates the unknowns.

Step 4 _____ the system of equations.

Step 5 _____ to the problem. Does it seem reasonable?

Step 6 _____ the answer in the words of the original problem.

EXAMPLE 5 Using a Linear System to Solve an Application

Salaries for the same position can vary depending on the location. In 2010, the average of the salaries for the position of Accountant I in San Diego, California, and Salt Lake City, Utah, was $45,091.50. The salary in San Diego, however, exceeded the salary in Salt Lake City by $5231. Determine the salary for the Accountant I position in San Diego and in Salt Lake City. (*Source:* www.salary.com)

Solving Linear Systems with Three Unknowns (Variables)

EXAMPLE 6 Solving a System of Three Equations with Three Variables
Solve the system.
$$3x + 9y + 6z = 3$$
$$2x + y - z = 2$$
$$x + y + z = 2$$

EXAMPLE 7 Solving a System of Two Equations with Three Variables
Solve the system.
$$x + 2y + z = 4$$
$$3x - y - 4z = -9$$

Using Systems of Equations to Model Data

EXAMPLE 8 Using Modeling to Find an Equation through Three Points
Find the equation of the parabola $y = ax^2 + bx + c$ that passes through the points $(2, 4)$, $(-1, 1)$, and $(-2, 5)$.

EXAMPLE 9 Solving an Application Using a System of Three Equations
An animal feed is made from three ingredients: corn, soybeans, and cottonseed. One unit of each ingredient provides units of protein, fat, and fiber as shown in the table. How many units of each ingredient should be used to make a feed that contains 22 units of protein, 28 units of fat, and 18 units of fiber?

	Corn	Soybeans	Cottonseed	Total
Protein	0.25	0.4	0.2	22
Fat	0.4	0.2	0.3	28
Fiber	0.3	0.2	0.1	18

5.2 Matrix Solution of Linear Systems
■ The Gauss-Jordan Method ■ Special Systems

Key Terms: matrix (matrices), element (of a matrix), augmented matrix, dimension (of a matrix)

The Gauss-Jordan Method

Matrix Row Transformations

For any augmented matrix of a system of linear equations, the following row transformations will result in the matrix of an equivalent system.

1. Interchange any two rows.
2. Multiply or divide the elements of any row by a nonzero real number.
3. Replace any row of the matrix by the sum of the elements of that row and a multiple of the elements of another row.

Using the Gauss-Jordan Method to Put a Matrix into Diagonal Form

Step 1 _____

Step 2 _____

_____

Step 3 _____

Step 4 _____

_____

Step 5 _____

The Gauss-Jordan method proceeds column by column, from _____ to _____.

EXAMPLE 1 Using the Gauss-Jordan Method
Solve the system.
$$3x - 4y = 1$$
$$5x + 2y = 19$$

EXAMPLE 2 Using the Gauss-Jordan Method
Solve the system.
$$x - y + 5z = -6$$
$$3x + 3y - z = 10$$
$$x + 3y + 2z = 5$$

Reflect: *In your own words, describe the step-by-step process of solving a linear system using the Gauss-Jordan Method.*

Special Systems

EXAMPLE 3 Solving an Inconsistent System
Use the Gauss-Jordan method to solve the system.

$$x + y = 2$$
$$2x + 2y = 5$$

EXAMPLE 4 Solving a System with Infinitely Many Solutions
Use the Gauss-Jordan method to solve the system.

$$2x - 5y + 3z = 1$$
$$x - 2y - 2z = 8$$

Summary of Possible Cases

When matrix methods are used to solve a system of linear equations and the resulting matrix is written in diagonal form:

1. If the number of rows with nonzero elements to the left of the vertical line is equal to the number of variables in the system, then the system has a _____ solution. **See Examples 1 and 2.**
2. If one of the rows has the form $[0\ 0\ \cdots\ 0\,|\,a]$ with $a \neq 0$, then the system has _____ solution. **See Example 3.**
3. If there are fewer rows in the matrix containing nonzero elements than the number of variables, then the system has either _____ solution or _____ _____ solutions. If there are _____ _____ solutions, give the solutions in terms of one or more arbitrary variables. **See Example 4.**

5.3 Determinant Solution of Linear Systems
■ Determinants ■ Cofactors ■ Evaluating $n \times n$ Determinants
■ Determinant Theorems ■ Cramer's Rule

Key Terms: determinant, minor, cofactor, expansion by a row or column, Cramer's rule

Determinants

Determinant of a 2×2 Matrix

If $A = \begin{bmatrix} a_{11} & a_{12} \\ a_{21} & a_{22} \end{bmatrix}$, then $|A| = \begin{vmatrix} a_{11} & a_{12} \\ a_{21} & a_{22} \end{vmatrix} = a_{11}a_{22} - a_{21}a_{12}$.

Matrices are enclosed by _____ _____, while determinants are denoted by _____ _____.

EXAMPLE 1 Evaluating a 2×2 Determinant

Let $A = \begin{bmatrix} -3 & 4 \\ 6 & 8 \end{bmatrix}$. Find $|A|$.

Determinant of a 3×3 Matrix

If $A = \begin{bmatrix} a_{11} & a_{12} & a_{13} \\ a_{21} & a_{22} & a_{23} \\ a_{31} & a_{32} & a_{33} \end{bmatrix}$, then the determinant of A, symbolized $|A|$, is defined as follows.

$$|A| = \begin{vmatrix} a_{11} & a_{12} & a_{13} \\ a_{21} & a_{22} & a_{23} \\ a_{31} & a_{32} & a_{33} \end{vmatrix} = (a_{11}a_{22}a_{33} + a_{12}a_{23}a_{31} + a_{13}a_{21}a_{32}) - (a_{31}a_{22}a_{13} + a_{32}a_{23}a_{11} + a_{33}a_{21}a_{12})$$

Cofactors

Cofactor

Let M_{ij} be the minor for element a_{ij} in an $n \times n$ matrix. The **cofactor** of a_{ij}, written A_{ij}, is defined as follows.

$$A_{ij} = (-1)^{i+j} \cdot M_{ij}$$

EXAMPLE 2 Finding Cofactors of Elements

Find the cofactor of each of the following elements of the given matrix.

$$\begin{bmatrix} 6 & 2 & 4 \\ 8 & 9 & 3 \\ 1 & 2 & 0 \end{bmatrix}$$

(a) 6   (b) 3   (c) 8

Evaluating $n \times n$ Determinants

Finding the Determinant of a Matrix

Multiply each element in any row or column of the matrix by its cofactor. The sum of these products gives the value of the determinant.

EXAMPLE 3 Evaluating a 3×3 Determinant

Evaluate $\begin{vmatrix} 2 & -3 & -2 \\ -1 & -4 & -3 \\ -1 & 0 & 2 \end{vmatrix}$, expanding by the second column.

Determinant Theorems

Determinant Theorems

1. If every element in a row (or column) of matrix A is 0, then $|A| =$ _____.
2. If the rows of matrix A are the corresponding columns of matrix B, then $|B| =$ _____.
3. If any two rows (or columns) of matrix A are interchanged to form matrix B, then $|B| =$ _____.
4. Suppose matrix B is formed by multiplying every element of a row (or column) of matrix A by the real number k. Then $|B| =$ _____.
5. If two rows (or columns) of matrix A are identical, then $|A| =$ _____.
6. Changing a row (or column) of a matrix by adding to it a constant times another row (or column) does not change the determinant of the matrix.
7. If the matrix A is in triangular form, having zeros either above or below the main diagonal, then $|A|$ is the product of the elements on the _____ _____ of A.

EXAMPLE 4 Using the Determinant Theorems

Use the determinant theorems to find the value of each determinant.

(a) $\begin{vmatrix} -2 & 4 & 2 \\ 6 & 7 & 3 \\ 0 & 16 & 8 \end{vmatrix}$

(b) $\begin{vmatrix} 3 & -7 & 4 & 10 \\ 0 & 1 & 8 & 3 \\ 0 & 0 & -5 & 2 \\ 0 & 0 & 0 & 6 \end{vmatrix}$

Cramer's Rule

Cramer's Rule for Two Equations in Two Variables

Given the system
$$a_1 x + b_1 y = c_1$$
$$a_2 x + b_2 y = c_2,$$

if $D \neq 0$, then the system has the unique solution

$$x = \frac{D_x}{D} \quad \text{and} \quad y = \frac{D_y}{D},$$

where $D = \begin{vmatrix} a_1 & b_1 \\ a_2 & b_2 \end{vmatrix}$, $D_x = \begin{vmatrix} c_1 & b_1 \\ c_2 & b_2 \end{vmatrix}$, and $D_y = \begin{vmatrix} a_1 & c_1 \\ a_2 & c_2 \end{vmatrix}$.

Cramer's Rule does not apply if _____.

EXAMPLE 5 Applying Cramer's Rule to a 2×2 System

Use Cramer's Rule to solve the system.

$$5x + 7y = -1$$
$$6x + 8y = 1$$

General Form of Cramer's Rule

Let an $n \times n$ system have linear equations of the form

$$a_1x_1 + a_2x_2 + a_3x_3 + \cdots + a_nx_n = b.$$

Define D as the determinant of the $n \times n$ matrix of all coefficients of the variables. Define D_{x_1} as the determinant obtained from D by replacing the entries in column 1 of D with the constants of the system. Define D_{x_i} as the determinant obtained from D by replacing the entries in column i with the constants of the system. If $D \neq 0$, the unique solution of the system is

$$x_1 = \frac{D_{x_1}}{D}, \quad x_2 = \frac{D_{x_2}}{D}, \quad x_3 = \frac{D_{x_3}}{D}, \quad \ldots, \quad x_n = \frac{D_{x_n}}{D}.$$

EXAMPLE 6 Applying Cramer's Rule to a 3×3 System

Use Cramer's Rule to solve the system.

$$x + y - z + 2 = 0$$
$$2x - y + z + 5 = 0$$
$$x - 2y + 3z - 4 = 0$$

EXAMPLE 7 Showing That Cramer's Rule Does Not Apply

Show that Cramer's Rule does not apply to the following system.

$$2x - 3y + 4z = 10$$
$$6x - 9y + 12z = 24$$
$$x + 2y - 3z = 5$$

When $D = 0$, the system is either _____ or has _____ _____ _____.

5.4 Partial Fractions

- Decomposition of Rational Expressions ■ Distinct Linear Factors
- Repeated Linear Factors ■ Distinct Linear and Quadratic Factors
- Repeated Quadratic Factors

Key Terms: partial fraction decomposition, partial fraction

Decomposition of Rational Expressions

A special type of sum involving rational expressions is a **partial fraction decomposition** - each term in the sum is a **partial fraction**.

$$\xrightarrow{}$$
$$\frac{2}{x+1} + \frac{3}{x} = \frac{5x+3}{x(x+1)}$$
$$\xleftarrow{}$$

Partial Fraction Decomposition of $\dfrac{f(x)}{g(x)}$

To form a partial fraction decomposition of a rational expression, follow these steps.

Step 1 If $\dfrac{f(x)}{g(x)}$ is not a proper fraction (a fraction with numerator of lesser degree than the denominator), divide $f(x)$ by $g(x)$. For example,

$$\frac{x^4 - 3x^3 + x^2 + 5x}{x^2 + 3} = \underline{} + \frac{14x+6}{x^2+3}$$

Then apply the following steps to the remainder, which is a proper fraction.

Step 2 _____ the denominator $g(x)$ completely into factors of the form $(ax+b)^m$ or $(cx^2+dx+e)^n$ where cx^2+dx+e is irreducible and m and n are positive integers.

Step 3 (a) For each distinct _____ factor $(ax+b)$, the decomposition must include the term $\dfrac{A}{ax+b}$.

(b) For each _____ linear factor $(ax+b)^m$, the decomposition must include the terms

$$\frac{A_1}{ax+b} + \frac{A_2}{(ax+b)^2} + \cdots + \frac{A_m}{(ax+b)^m}.$$

Step 4 (a) For each distinct _____ factor (cx^2+dx+e), the decomposition must include the term $\dfrac{Bx+C}{cx^2+dx+e}$.

(b) For each _____ quadratic factor $(cx^2+dx+e)^n$, the decomposition must include the terms
$$\dfrac{B_1x+C_1}{cx^2+dx+e}+\dfrac{B_2x+C_2}{(cx^2+dx+e)^2}+\cdots+\dfrac{B_nx+C_n}{(cx^2+dx+e)^n}.$$

Step 5 Use algebraic techniques to solve for the constants in the numerators of the decomposition.

Distinct Linear Factors

EXAMPLE 1 Finding a Partial Fraction Decomposition

Find the partial fraction decomposition of $\dfrac{2x^4-8x^2+5x-2}{x^3-4x}$.

Repeated Linear Factors

EXAMPLE 2 Finding a Partial Fraction Decomposition

Find the partial fraction decomposition of $\dfrac{2x}{(x-1)^3}$.

Distinct Linear and Quadratic Factors

EXAMPLE 3 Finding a Partial Fraction Decomposition

Find the partial fraction decomposition of $\dfrac{x^2+3x-1}{(x+1)(x^2+2)}$.

Repeated Quadratic Factors

EXAMPLE 4 Finding a Partial Fraction Decomposition

Find the partial fraction decomposition of $\dfrac{2x}{(x^2+1)^2(x-1)}$.

Reflect: *If the denominator of a rational expression consists of an irreducible quadratic factor to the m power and a linear factor to the n power, how many terms are in its partial fraction decomposition? How many constants must be determined?*

Techniques for Decomposition into Partial Fractions

Method 1 For Linear Factors

Step 1 Multiply each side of the resulting rational equation by the _____ _____.

Step 2 Substitute the zero of each factor in the resulting equation. For repeated linear factors, substitute as many other numbers as necessary to find all the constants in the numerators. The number of substitutions required will _____ the number of constants A, B, ...

Method 2 For Quadratic Factors

Step 1 Multiply each side of the resulting rational equation by the _____ _____.

Step 2 Collect like terms on the right side of the equation.

Step 3 Equate the _____ of like terms to get a system of equations.

Step 4 Solve the system to find the constants in the numerators.

5.5 Nonlinear Systems of Equations
- **Solving Nonlinear Systems with Real Solutions**
- **Solving Nonlinear Systems with Nonreal Complex Solutions**
- **Applying Nonlinear Systems**

Key Terms: nonlinear system

__Solving Nonlinear Systems with Real Solutions__

A system of equations in which at least one equation is *not* linear is a _____ _____.

The substitution method works well for solving many such systems, particularly when one of the equations is linear, as in the next example.

EXAMPLE 1 Solving a Nonlinear System by Substitution
Solve the system.
$$x^2 - y = 4$$
$$x + y = -2$$

CHAPTER 5 Systems and Matrices

Nonlinear systems where both variables are squared in both equations are best solved by elimination, as shown in the next example.

EXAMPLE 2 Solving a Nonlinear System by Elimination
Solve the system.
$$x^2 + y^2 = 4$$
$$2x^2 - y^2 = 8$$

EXAMPLE 3 Solving a Nonlinear System by a Combination of Methods
Solve the system.
$$x^2 + 3xy + y^2 = 22$$
$$x^2 - xy + y^2 = 6$$

EXAMPLE 4 Solving a Nonlinear System with an Absolute Value Equation
Solve the system.
$$x^2 + y^2 = 16$$
$$|x| + y = 4$$

Solving Nonlinear Systems with Nonreal Complex Solutions

EXAMPLE 5 Solving a Nonlinear System with Nonreal Complex Numbers as Solutions
Solve the system.
$$x^2 + y^2 = 5$$
$$4x^2 + 3y^2 = 11$$

Reflect: *In Example 5 there are four solutions to the system, but there are no points of intersection of the graphs of the equations. If a nonlinear system has nonreal complex numbers as components of its solutions, will they appear as intersection points of the graphs?*

Applying Nonlinear Systems

EXAMPLE 6 Using a Nonlinear System to Find the Dimensions of a Box

A box with an open top has a square base and four sides of equal height. The volume of the box is 75 in.^{3}, and the surface area is 85 in.^{2}. What are the dimensions of the box?

Reflect: *In this chapter, we have solved systems of equations, both linear and nonlinear. How do you determine whether a system is linear or nonlinear?*

Reflect: *Describe methods for solving linear systems. Describe methods for solving nonlinear systems.*

5.6 Systems of Inequalities and Linear Programming
■ Solving Linear Inequalities ■ Solving Systems of Inequalities ■ Linear Programming

Key Terms: half-plane, boundary, linear inequality in two variables, system of inequalities, linear programming, constraints, objective function, region of feasible solutions, vertex (corner point)

Solving Linear Inequalities

Linear Inequality in Two Variables
A **linear inequality in two variables** is an inequality of the form
$$Ax + By \leq C,$$
where A, B, and C are real numbers, with A and B not both equal to 0. (The symbol $\leq$ could be replaced with $\geq$, $<$, or $>$.)

EXAMPLE 1 Graphing a Linear Inequality
Graph $3x - 2y \leq 6$.

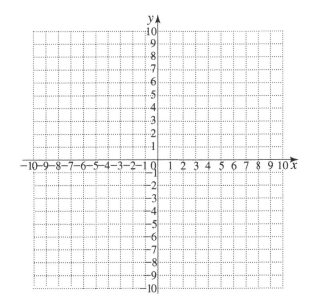

A linear inequality must be in slope-intercept form (solved for y) to determine, from the presence of a < symbol or a > symbol, whether to shade the lower or upper half-plane.

EXAMPLE 2 Graphing a Linear Inequality

Graph $x + 4y > 4$.

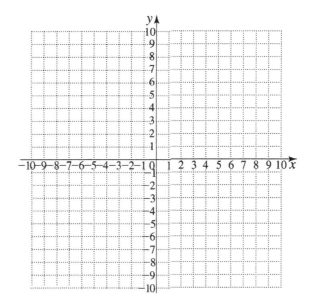

Reflect: How do you determine whether the boundary of the graph of an inequality is solid or dashed?

Graphing an Inequality in Two Variables

Method 1 If the inequality is or can be solved for y, then the following hold.

- The graph of $y < f(x)$ consists of all the points that are _____ the graph of $y = f(x)$.
- The graph of $y > f(x)$ consists of all the points that are _____ the graph of $y = f(x)$.

Method 2 If the inequality is not or cannot be solved for y, choose a test point not on the boundary.

- If the test point satisfies the inequality, the graph includes all points on the _____ side of the boundary as the test point,
- If the test point does not satisfy the inequality, the graph includes all points on the _____ side of the boundary.

Solving Systems of Inequalities

EXAMPLE 3 Graphing Systems of Inequalities
Graph the solution set of each system.

(a) $x > 6 - 2y$
$x^2 < 2y$

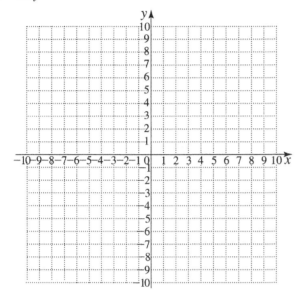

(b) $|x| \leq 3$
$y \leq 0$
$y \geq |x| + 1$

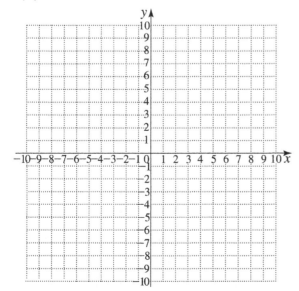

MN-256   CHAPTER 5 Systems and Matrices

Reflect: *Finish this sentence. The shaded region, or solution set, of a system of inequalities consists of points that . . .*

Linear Programming

EXAMPLE 4 Finding a Maximum Profit

The Charlson Company makes two products: MP3 players and DVD players. Each MP3 player gives a profit of $30, while each DVD player $70 profit. The company must manufacture at least 10 MP3 players per day to satisfy one of tis customers, but it can manufacture no more than 50 per day because of production restrictions. The number of DVD players produced cannot exceed 60 per day, and the number of MP3 players cannot exceed the number of DVD players. How many of each should the company manufacture to obtain maximum profit?

Fundamental Theorem of Linear Programming

If an optimal value for a linear programming problem exists, it occurs at a _____ of the region of feasible solutions.

Solving a Linear Programming Problem

Step 1 _____

Step 2 _____

Step 3 _____

Step 4 _____

Step 5 _____

EXAMPLE 5 Finding a Minimum Cost

Robin takes multivitamins each day. She wants at least 16 units of vitamin A, at least 5 units of vitamin B_1, and at least 20 units of vitamin C. Capsules, costing $0.10 each, contain 8 units of A, 1 of B_1, and 2 of C. Chewable tablets, costing $0.20 each, contain 2 units of A, 1 of B_1, and 7 of C. How many of each should she take each day to minimize her cost and yet fulfill her daily requirements?

5.7 Properties of Matrices
■ Basic Definitions ■ Adding Matrices ■ Special Matrices ■ Subtracting Matrices
■ Multiplying Matrices ■ Applying Matrix Algebra

Key Terms: square matrix, row matrix, column matrix, zero matrix, additive inverse (negative) of a matrix, scalar

Basic Definitions

Certain matrices have special names: an $n \times n$ matrix is a _____ **matrix** because the number of rows is equal to the number of columns. A matrix with just one row is a _____ **matrix,** and a matrix with just one column is a _____ **matrix.**

Two matrices are equal if they have the same _____ and if corresponding elements, position by position, are _____.

EXAMPLE 1 Finding Values to Make Two Matrices Equal
Find the values of the variables for which each statement is true, if possible.

(a) $\begin{bmatrix} 2 & 1 \\ p & q \end{bmatrix} = \begin{bmatrix} x & y \\ -1 & 0 \end{bmatrix}$

(b) $\begin{bmatrix} x \\ y \end{bmatrix} = \begin{bmatrix} 1 \\ 4 \\ 0 \end{bmatrix}$

Adding Matrices

Addition of Matrices
To add two matrices of the same dimension, add _____ elements. *Only matrices of the same _____ can be added.*

EXAMPLE 2 Adding Matrices
Find each sum, if possible.

(a) $\begin{bmatrix} 5 & -6 \\ 8 & 9 \end{bmatrix} + \begin{bmatrix} -4 & 6 \\ 8 & -3 \end{bmatrix}$

(b) $\begin{bmatrix} 2 \\ 5 \\ 8 \end{bmatrix} + \begin{bmatrix} -6 \\ 3 \\ 12 \end{bmatrix}$

(c) $A + B$, if $A = \begin{bmatrix} 5 & 8 \\ 6 & 2 \end{bmatrix}$ and $B = \begin{bmatrix} 3 & 9 & 1 \\ 4 & 2 & 5 \end{bmatrix}$

Special Matrices

A matrix containing only _____ elements is a **zero matrix**.

Given a matrix A, there is a matrix $-A$ such that $A + (-A) = O$. The matrix $-A$ has as elements the _____ inverses of the elements of A.

Subtracting Matrices

Subtraction of Matrices
If A and B are two matrices of the same dimension, then

$$A - B = A + (-B).$$

EXAMPLE 3 Subtracting Matrices
Find each difference, if possible.

(a) $\begin{bmatrix} -5 & 6 \\ 2 & 4 \end{bmatrix} - \begin{bmatrix} -3 & 2 \\ 5 & -8 \end{bmatrix}$

(b) $\begin{bmatrix} 8 & 6 & -4 \end{bmatrix} - \begin{bmatrix} 3 & 5 & -8 \end{bmatrix}$

(c) $A - B$, if $A = \begin{bmatrix} -2 & 5 \\ 0 & 1 \end{bmatrix}$ and $B = \begin{bmatrix} 3 \\ 5 \end{bmatrix}$

Multiplying Matrices

In work with matrices, a real number is called a **scalar** to distinguish it from a matrix. The product of a scalar k and a matrix X is the matrix kX, each of whose elements is k times the corresponding element of X.

EXAMPLE 4 Multiplying Matrices by Scalars
Find each product.

(a) $5\begin{bmatrix} 2 & -3 \\ 0 & 4 \end{bmatrix}$

(b) $\dfrac{3}{4}\begin{bmatrix} 20 & 36 \\ 12 & -16 \end{bmatrix}$.

Properties of Scalar Multiplication
Let A and B be matrices of the same dimension, and let c and d be scalars. Then these properties hold.

$(c+d)A =$ _____ $(cA)d =$ _____

$c(A+B) =$ _____ $(cd)A =$ _____

Matrix Multiplication
The number of columns of an $m \times n$ matrix A is the same as the number of rows of an $n \times p$ matrix B (i.e. both n). The element c_{ij} of the product matrix $C = AB$ is found as follows.

$$c_{ij} = a_{i1}b_{1j} + a_{i2}b_{2j} + \cdots + a_{in}b_{nj}$$

Matrix AB will be an $m \times p$ matrix.

CHAPTER 5 Systems and Matrices

EXAMPLE 5 Deciding Whether Two Matrices Can Be Multiplied
Suppose A is a 3×2 matrix, while B is a 2×4 matrix.

(a) Can the product AB be calculated?

(b) If AB can be calculated, what is its dimension?

(c) Can BA be calculated?

(d) If BA can be calculated, what is its dimension?

EXAMPLE 6 Multiplying Matrices

Let $A = \begin{bmatrix} 1 & -3 \\ 7 & 2 \end{bmatrix}$ and $B = \begin{bmatrix} 1 & 0 & -1 & 2 \\ 3 & 1 & 4 & -1 \end{bmatrix}$. Find each product, if possible.

(a) AB

(b) BA

EXAMPLE 7 Multiplying Square Matrices in Different Orders

Let $A = \begin{bmatrix} 1 & 3 \\ -2 & 5 \end{bmatrix}$ and $B = \begin{bmatrix} -2 & 7 \\ 0 & 2 \end{bmatrix}$. Find each product.

(a) AB

(b) BA

Reflect: Is matrix multiplication commutative?

Properties of Matrix Multiplication
If A, B, and C are matrices such that all the following products and sums exist, then these properties hold.

$(AB)C = $ _____, $A(B+C) = $ _____, $(B+C)A = $ _____

Reflect: What are the names of the various matrix multiplication properties?

Applying Matrix Algebra

EXAMPLE 8 Using Matrix Multiplication to Model Plans for a Subdivision

A contractor builds three kinds of houses, models A, B, and C, with a choice of two styles, colonial or ranch. Matrix P below shows the number of each kind of house the contractor is planning to build for a new 100-home subdivision. The amounts for each of the main materials used depend on the style of the house. These amounts are shown in matrix Q, while matrix R gives the cost in dollars for each kind of material. Concrete is measured here in cubic yards, lumber in 1000 board feet, brick in 1000s, and shingles in 100 square feet

$$\begin{array}{c} \text{Colonial Ranch} \\ \begin{array}{c}\text{Model A}\\ \text{Model B}\\ \text{Model C}\end{array}\begin{bmatrix} 0 & 30 \\ 10 & 20 \\ 20 & 20 \end{bmatrix} = P \end{array}$$

$$\begin{array}{c} \text{Concrete Lumber Brick Shingles} \\ \begin{array}{c}\text{Colonial}\\ \text{Ranch}\end{array}\begin{bmatrix} 10 & 2 & 0 & 2 \\ 50 & 1 & 20 & 2 \end{bmatrix} = Q \end{array}$$

$$\begin{array}{c} \text{Cost per Unit} \\ \begin{array}{c}\text{Concrete}\\ \text{Lumber}\\ \text{Brick}\\ \text{Shingles}\end{array}\begin{bmatrix} 20 \\ 180 \\ 60 \\ 25 \end{bmatrix} = R \end{array}$$

(a) What is the total cost of materials for all houses of each model?

(b) How much of each of the four kinds of material must be ordered?

(c) What is the total cost of the materials?

5.8 Matrix Inverses
■ Identity Matrices ■ Multiplicative Inverses ■ Solving Systems Using Inverse Matrices

Key Terms: identity matrix, multiplicative inverse (of a matrix)

Identity Matrices

By the identity property for real numbers,

$$a \cdot 1 = a \quad \text{and} \quad 1 \cdot a = a$$

for any real number a. If there is to be a multiplicative **identity matrix I,** such that

$$AI = A \quad \text{and} \quad IA = A,$$

for any matrix A, then A and I must be square matrices of the same dimension.

2×2 Identity Matrix
If I_2 represents the 2×2 identity matrix, then

$$I_2 = \begin{bmatrix} 1 & 0 \\ 0 & 1 \end{bmatrix}.$$

Generalizing, there is an $n \times n$ identity matrix for every $n \times n$ square matrix. The $n \times n$ identity matrix has _____ on the main diagonal and _____ elsewhere.

$n \times n$ Identity Matrix
The $n \times n$ identity matrix is I_n, where

$$I_n = \begin{bmatrix} 1 & 0 & \cdots & 0 \\ 0 & 1 & \cdots & 0 \\ \vdots & \vdots & a_{ij} & \vdots \\ 0 & 0 & \cdots & 1 \end{bmatrix}$$

The element _____ when $i = j$ (the diagonal elements), and _____ otherwise.

Section 5.8 MyNotes

EXAMPLE 1 Verifying the Identity Property of I_3

Let $A = \begin{bmatrix} -2 & 4 & 0 \\ 3 & 5 & 9 \\ 0 & 8 & -6 \end{bmatrix}$. Give the 3×3 identity matrix I_3 and show that $AI_3 = A$.

Reflect: What is the product $I_n I_n$?

Multiplicative Inverses

For every nonzero real number a, there is a multiplicative inverse $\dfrac{1}{a}$ that satisfies both of the following.

$$a \cdot \dfrac{1}{a} = 1 \quad \text{and} \quad \dfrac{1}{a} \cdot a = 1$$

In a similar way, if A is an $n \times n$ matrix, then its **multiplicative inverse,** written A^{-1}, must satisfy both of the following.

$$AA^{-1} = I_n \quad \text{and} \quad A^{-1}A = I_n$$

This means that only a _____ matrix can have a multiplicative inverse.

Finding an Inverse Matrix
To obtain A^{-1} for any $n \times n$ matrix A for which A^{-1} exists, follow these steps.

Step 1 _____

Step 2 _____

Step 3 _____

EXAMPLE 2 Finding the Inverse of a 3×3 Matrix

Find A^{-1} if $A = \begin{bmatrix} 1 & 0 & 1 \\ 2 & -2 & -1 \\ 3 & 0 & 0 \end{bmatrix}$.

EXAMPLE 3 Identifying a Matrix with No Inverse

Find A^{-1}, if possible, given that $A = \begin{bmatrix} 2 & -4 \\ 1 & -2 \end{bmatrix}$.

Solving Systems Using Inverse Matrices

Solution of the Matrix Equation $AX = B$

If A is an $n \times n$ matrix with inverse A^{-1}, X is an $n \times 1$ matrix of variables, and B is an $n \times 1$ matrix, then the matrix equation

$$AX = B$$

has the solution $\qquad X = \underline{}.$

Reflect: If a and b are numbers, what is the solution of $ax = b$?

EXAMPLE 4 Solving Systems of Equations Using Matrix Inverses

Use the inverse of the coefficient matrix to solve each system.

(a) $2x - 3y = 4$
    $x + 5y = 2$

(b) $x + z = -1$
    $2x - 2y - z = 5$
    $3x = 6$

Name: Date:
Instructor: Section:

Chapter 1 Equations and Inequalities

1.1R Properties of Real Numbers, Evaluating Exponential Expressions, Order of Operations, Simplifying Expressions, Operations on Real Numbers

Key Terms

Use the vocabulary terms listed below to complete each statement in exercises 1–16.

- **identity element for addition**
- **identity element for multiplication**

exponent	**base**	**exponential expression**
term	**numerical coefficient**	**like terms**
sum	**addends**	**minuend**
subtrahend	**difference**	**product**
quotient	**reciprocals**	

1. When the _____, which is 0, is added to a number, the number is unchanged.

2. When a number is multiplied by the _____, which is 1, the number is unchanged.

3. A number written with an exponent is an _____.

4. The _____ is the number that is a repeated factor when written with an exponent.

5. An _____ is a number that indicates how many times a factor is repeated.

6. In the term $4x^2$, "4" is the _____.

7. A number, a variable, or a product or quotient of a number and one or more variables raised to powers is called a _____.

8. Terms with exactly the same variables, including the same exponents, are called _____.

9. The answer to an addition problem is called the _____.

10. In an addition problem, the numbers being added are the _____.

CHAPTER 1 Equations and Inequalities

Name: _____ Date: _____
Instructor: _____ Section: _____

11. The number from which another number is being subtracted is called the _____.

12. The _____ is the number being subtracted.

13. The answer to a subtraction problem is called the _____.

14. The answer to a division problem is called the _____.

15. Pairs of numbers whose product is 1 are called _____.

16. The answer to a multiplication problem is called the _____.

Guided Examples

Review these examples:

1. Use a commutative property to complete each statement.

   a. $-7+6=6+$ _____

   Using the commutative property of addition,
   $-7+6=6+(-7)$

   b. $(-3)5=$ _____ (-3)

   Using the commutative property of multiplication,
   $(-3)5=5(-3)$

2. Use an associative property to complete each statement.

   a. $-5+(3+7)=(-5+$ _____ $)+7$

   Using the associative property of addition,
   $-5+(3+7)=(-5+3)+7$

   b. $[4 \cdot (-9)] \cdot 2 = 4 \cdot$ _____

   Using the associative property of multiplication,
   $[4 \cdot (-9)] \cdot 2 = 4 \cdot [(-9) \cdot 2]$

Now Try:

1. Use a commutative property to complete each statement.

   a. $-12+8=8+$ _____

   _____

   b. $(-4)2=$ _____ (-4)

   _____

2. Use an associative property to complete each statement.

   a. $-8+(4+6)=(-8+$ ___$)+6$

   _____

   b. $[8 \cdot (-3)] \cdot 4 = 8 \cdot$ _____

   _____

Name: Date:
Instructor: Section:

3. Decide whether each statement is an example of a commutative property, an associative property, or both.

   a. $(5+9)+11 = 5+(9+11)$

   The order of the three numbers is the same, but the change is in grouping. This is an example of the associative property.

   b. $7 \cdot (9 \cdot 11) = 7 \cdot (11 \cdot 9)$

   The only change involves the order of the number, so this is an example of the commutative property.

   c. $(12+3)+6 = 12+(6+3)$

   Both the order and the grouping are changed. This is an example of both the associative and commutative properties.

4. Find each sum or product.

   a. $54+21+3+17+29$

   $54+21+3+17+29$
   $= 54+(21+29)+(3+17)$
   $= 54+50+20$
   $= 124$

   b. $50(43)(4)$

   $50(43)(4) = 50(4)(43)$
   $\quad = 200(43)$
   $\quad = 8600$

5. Use an identity property to complete each statement.

   a. $-6 + \underline{\quad} = -6$

   Use the identity property for addition.
   $-6 + 0 = -6$

3. Decide whether each statement is an example of a commutative property, an associative property, or both.

   a. $(13+8)+25 = 13+(8+25)$

   _____

   b. $4 \cdot (15 \cdot 30) = 4 \cdot (30 \cdot 15)$

   _____

   c. $(21+19)+4 = 21+(4+19)$

   _____

4. Find each sum or product.

   a. $48+15+12+24+8$

   _____

   b. $40(63)(5)$

   _____

5. Use an identity property to complete each statement.

   a. $8 + \underline{\quad} = 8$

   _____

CHAPTER 1 Equations and Inequalities

Name: Date:
Instructor: Section:

b. ____ $\cdot \dfrac{1}{6} = \dfrac{1}{6}$

Use the identity property for multiplication.
$1 \cdot \dfrac{1}{6} = \dfrac{1}{6}$

b. $-9 \cdot$ ____ $= -9$

____

6. a. Write $\dfrac{56}{35}$ in lowest terms.

$\dfrac{56}{35} = \dfrac{8 \cdot 7}{5 \cdot 7}$

$= \dfrac{8}{5} \cdot \dfrac{7}{7}$

$= \dfrac{8}{5} \cdot 1$

$= \dfrac{8}{5}$

6. a. Write $\dfrac{49}{63}$ in lowest terms.

____

b. Perform the operation: $\dfrac{5}{6} - \dfrac{7}{18}$

$\dfrac{5}{6} - \dfrac{7}{18} = \dfrac{5}{6} \cdot 1 - \dfrac{7}{18}$

$= \dfrac{5}{6} \cdot \dfrac{3}{3} - \dfrac{7}{18}$

$= \dfrac{15}{18} - \dfrac{7}{18}$

$= \dfrac{8}{18}$

$= \dfrac{4}{9}$

b. Perform the operation:
$\dfrac{3}{7} + \dfrac{5}{21}$

____

7. Use an inverse property to complete each statement.

a. ____ $+ \dfrac{1}{4} = 0$

Use the inverse property of addition.
$-\dfrac{1}{4} + \dfrac{1}{4} = 0$

b. $5 +$ ____ $= 0$

Use the inverse property of addition.
$5 + (-5) = 0$

7. Use an inverse property to complete each statement.

a. $-11 +$ ____ $= 0$

____

b. $8 +$ ____ $= 0$

____

Copyright © 2015 Pearson Education, Inc.

Name: Date:
Instructor: Section:

c. $-0.4 + \frac{2}{5} = $ _____

Use the inverse property of addition.
$-0.4 + \frac{2}{5} = 0$

d. _____ $\cdot \frac{6}{7} = 1$

Use the inverse property of multiplication.
$\frac{7}{6} \cdot \frac{6}{7} = 1$

e. $-9(\underline{}) = 1$

Use the inverse property of multiplication.
$-9\left(-\frac{1}{9}\right) = 1$

f. $8(0.125) = $ _____

Use the inverse property of multiplication.
$8(0.125) = 1$

8. Use the distributive property to rewrite each expression.

a. $4(8+7)$

$4(8+7) = 4 \cdot 8 + 4 \cdot 7$
$ = 32 + 28$
$ = 60$

b. $5(3+x+m)$

$5(3+x+m) = 5 \cdot 3 + 5x + 5m$
$ = 15 + 5x + 5m$

c. $-8(x+5)$

$-8(x+5) = -8x + (-8)(5)$
$ = -8x + (-40)$
$ = -8x - 40$

c. $-0.8 + \frac{4}{5} = $ _____

_____

d. $\frac{8}{5} \cdot $ _____ $= 1$

_____

e. $-\frac{1}{10}(\underline{}) = 1$

_____

f. $0.8\left(\frac{5}{4}\right) = $ _____

_____

8. Use the distributive property to rewrite each expression.

a. $3(11+7)$

_____

b. $12(y+6+x)$

_____

c. $-13(x+6)$

_____

CHAPTER 1 Equations and Inequalities

Name: Date:
Instructor: Section:

d. $7(p-6)$

$7(p-6) = 7[p+(-6)]$
$\qquad = 7p + 7(-6)$
$\qquad = 7p - 42$

e. $3(4t+18x+7z)$

$3(4t+18x+7z)$
$= 3(4t) + 3(18x) + 3(7z)$
$= (3 \cdot 4)t + (3 \cdot 18)x + (3 \cdot 7)z$
$= 12t + 54x + 21z$

f. $4 \cdot 9 + 4 \cdot 2$

Use the distributive property in reverse.
$4 \cdot 9 + 4 \cdot 2 = 4(9+2)$
$\qquad\qquad = 4(11)$
$\qquad\qquad = 44$

g. $5a - 5b$

Use the distributive property in reverse.
$\quad 5a - 5b = 5(a-b)$

9. Write each expression without parentheses.

a. $-(5x+7)$

$-(5x+7) = -1 \cdot (5x+7)$
$\qquad\qquad = -1 \cdot 5x + (-1) \cdot 7$
$\qquad\qquad = -5x - 7$

b. $-(-8w-3)$

$-(-8w-3) = -1 \cdot (-8w-3)$
$\qquad\qquad = -1 \cdot (-8w) - 1 \cdot (-3)$
$\qquad\qquad = 8w + 3$

d. $17(x-6)$

_____

e. $6(5x+14y+21z)$

_____

f. $25 \cdot 9 + 25 \cdot 6$

_____

g. $23x - 23y$

_____

9. Write each expression without parentheses.

a. $-(3x+4)$

_____

b. $-(-10x-7)$

_____

Name: Date:
Instructor: Section:

c. $-(-p-5r+9x)$

$-(-p-5r+9x)$
$=-1\cdot(-1p-5r+9x)$
$=-1\cdot(-1p)-1\cdot(-5r)-1\cdot(9x)$
$=p+5r-9x$

10. Find the value of each exponential expression.

  a. 6^2

  6^2 means $6\cdot 6$, which equals 36.

  b. 5^3

  5^3 means $5\cdot 5\cdot 5$, which equals 125.

  c. 3^4

  3^4 means $3\cdot 3\cdot 3\cdot 3$, which equals 81.

  d. $\left(\dfrac{3}{5}\right)^3$

  $\left(\dfrac{3}{5}\right)^3$ means $\dfrac{3}{5}\cdot\dfrac{3}{5}\cdot\dfrac{3}{5}$, which equals $\dfrac{27}{125}$.

  e. $(0.9)^2$

  $(0.9)^2$ means $0.9(0.9)$, which equals 0.81.

11. Simplify.

  a. $-7(3)-(-5)(4)$

  $-7(3)-(-5)(4) = -21-(-20)$
  $= -21+20$
  $= -1$

c. $-(-4x-5y+z)$

_____

10. Find the value of each exponential expression.

  a. 7^2

  _____

  b. 4^3

  _____

  c. 8^4

  _____

  d. $\left(\dfrac{5}{6}\right)^2$

  _____

  e. $(0.6)^2$

  _____

11. Simplify.

  a. $-11(3)-(-7)(5)$

  _____

CHAPTER 1 Equations and Inequalities

Name: Date:
Instructor: Section:

b. $-8(-3)-5(-6)$

$$-8(-3)-5(-6) = 24-(-30)$$
$$= 24+30$$
$$= 54$$

c. $\dfrac{6(-4)-5(3)}{3(2-7)}$

$$\dfrac{6(-4)-5(3)}{3(2-7)} = \dfrac{-24-15}{3(-5)}$$
$$= \dfrac{-39}{-15}$$
$$= \dfrac{13}{5}$$

12. Combine like terms in each expression.

    a. $-7m+4m$

$$-7m+4m = (-7+4)m$$
$$= -3m$$

    b. $8r+5r+4r$

$$8r+5r+4r = (8+5+4)r$$
$$= 17r$$

    c. $9x+x$

$$9x+x = 9x+1x$$
$$= (9+1)x$$
$$= 10x$$

    d. $15x^2-8x^2$

$$15x^2-8x^2 = (15-8)x^2$$
$$= 7x^2$$

    e. $27x+39x^2$

These unlike terms cannot be combined.

b. $-4(-8)-(9)(2)$

_____

c. $\dfrac{-9(-3)+4(-8)}{-4(5-6)}$

_____

12. Combine like terms in each expression.

    a. $-12m+6m$

_____

    b. $14r+7r+2r$

_____

    c. $18x+x$

_____

    d. $17x^2-9x^2$

_____

    e. $32x+16x^2$

_____

Name: Date:
Instructor: Section:

13. Simplify each expression.

a. $15y + 3(5 + 4y)$

$$15y + 3(5 + 4y) = 15y + 3(5) + 3(4y)$$
$$= 15y + 15 + 12y$$
$$= 27y + 15$$

b. $8k - 5 - 4(7 - 3k)$

$$8k - 5 - 4(7 - 3k) = 8k - 5 - 4(7) - 4(-3k)$$
$$= 8k - 5 - 28 + 12k$$
$$= 20k - 33$$

c. $-(5 - r) + 13r$

$$-(5 - r) + 13r = -1(5 - r) + 13r$$
$$= -1(5) - 1(-r) + 13r$$
$$= -5 + r + 13r$$
$$= -5 + 14r$$

d. $-\dfrac{3}{5}(x - 10) - \dfrac{1}{10}x$

$$-\dfrac{3}{5}(x - 10) - \dfrac{1}{10}x = -\dfrac{3}{5}x - \dfrac{3}{5}(-10) - \dfrac{1}{10}x$$
$$= -\dfrac{3}{5}x + 6 - \dfrac{1}{10}x$$
$$= -\dfrac{6}{10}x + 6 - \dfrac{1}{10}x$$
$$= -\dfrac{7}{10}x + 6$$

14. Use a number line to find the sum $4 + 4$.

Step 1 Start at 0 and draw an arrow 4 units to the right.

Step 2 From the right end of that arrow, draw another arrow 4 units to the right.

The number below the end of this second arrow is 8, so $4 + 4 = 8$.

13. Simplify each expression.

a. $9y + 5(3 + 8y)$

_____

b. $7k - 9 - 5(3 - 6k)$

_____

c. $-(12 - r) + 11r$

_____

d. $-\dfrac{3}{4}(x - 8) - \dfrac{1}{2}x$

_____

14. Use a number line to find the sum $2 + 4$.

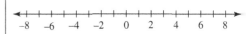

_____

CHAPTER 1 Equations and Inequalities

Name: Date:
Instructor: Section:

15. Use a number line to find the sum −3 + (−5).

    Step 1 Start at 0 and draw an arrow 3 units to the left.

    Step 2 From the left end of that arrow, draw another arrow 5 units to the left.

    The number below the end of this second arrow is −5, so −3 + (−5) = −8.

    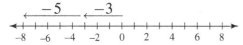

16. Use the number line to find the sum −3 + 4.

    Step 1 Start at 0 and draw an arrow 3 units to the left.

    Step 2 From the left end of that arrow, draw a second arrow 4 units to the right.

    The number below the end of this second arrow is 1, so −3 + 4 = 1.

    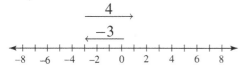

17. Find each sum.

    a. −3 + (−7)

    −3 + (−7) = −10

    b. −5 + (−16)

    −5 + (−16) = −21

    c. −18 + (−7)

    −18 + (−7) = −25

15. Use a number line to find the sum −4 + (−1).

    _____

16. Use the number line to find the sum 7 + (−4).

    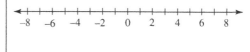

    _____

17. Find each sum.

    a. −8 + (−4)

    _____

    b. −17 + (−14)

    _____

    c. −10 + (−30)

    _____

Copyright © 2015 Pearson Education, Inc.

Name: Date:
Instructor: Section:

18. Find the sum $-10 + 6$.

Find the absolute value of each number.
$|-10| = 10$ and $|6| = 6$
Then find the difference between these absolute values: $10 - 6 = 4$. The sum will be negative since $|-10| > |6|$.
$-10 + 6 = -4$

18. Find the sum $-25 + 13$.

_____

19. Use a number line to find the difference $5 - 3$.

Step 1 Start at 0 and draw an arrow 5 units to the right.

Step 2 From the right end of that arrow, draw another arrow 3 units to the left.

The number below the end of this second arrow is 2, so $5 - 3 = 2$.

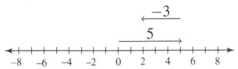

19. Use a number line to find the difference $3 - 1$.

_____

20. Subtract.

a. $13 - 4$

$13 - 4 = 13 + (-4) = 9$

b. $8 - 11$

$8 - 11 = 8 + (-11) = -3$

c. $-9 - 16$

$-9 - 16 = -9 + (-16) = -25$

d. $-6 - (-9)$

$-6 - (-9) = -6 + (9) = 3$

20. Subtract.

a. $12 - 9$

_____

b. $13 - 17$

_____

c. $-11 - 27$

_____

d. $-5 - (-7)$

_____

CHAPTER 1 Equations and Inequalities

Name: Date:
Instructor: Section:

e. $\dfrac{5}{6} - \left(-\dfrac{3}{7}\right)$

$\dfrac{5}{6} - \left(-\dfrac{3}{7}\right) = \dfrac{35}{42} - \left(-\dfrac{18}{42}\right)$
$= \dfrac{35}{42} + \dfrac{18}{42}$
$= \dfrac{53}{42}$

e. $\dfrac{5}{9} - \left(-\dfrac{4}{5}\right)$

_____

21. Find each product using the multiplication rule.

a. $9(-6)$

$9(-6) = -(9 \cdot 6) = -54$

b. $-8\left(\dfrac{1}{4}\right)$

$-8\left(\dfrac{1}{4}\right) = -2$

c. $-3.7(2.5)$

$-3.7(2.5) = -9.25$

21. Find each product using the multiplication rule.

a. $8(-7)$

_____

b. $-15\left(\dfrac{1}{5}\right)$

_____

c. $-9.8(4.6)$

_____

22. Find each product using the multiplication rule.

a. $-7(-3)$

$-7(-3) = 21$

b. $-8(-13)$

$-8(-13) = 104$

c. $-3(8)(-2)$

$-3(8)(-2) = -24(-2)$
$= 48$

d. $5(-4)(-7)$

$5(-4)(-7) = -20(-7)$
$= 140$

22. Find each product using the multiplication rule.

a. $-5(-6)$

_____

b. $-9(-15)$

_____

c. $-6(3)(-1)$

_____

d. $2(-9)(-5)$

_____

Name: Date:
Instructor: Section:

23. Find each quotient using the definition of division.

a. $\dfrac{14}{7}$

$\dfrac{14}{7} = 14 \cdot \dfrac{1}{7} = 2$

b. $\dfrac{3(-4)}{4}$

$\dfrac{3(-4)}{4} = -12 \cdot \dfrac{1}{4} = -3$

c. $\dfrac{-1.84}{-8}$

$\dfrac{-1.84}{-8} = -1.84 \cdot \left(-\dfrac{1}{8}\right) = 0.23$

d. $-\dfrac{4}{7} \div \left(-\dfrac{11}{2}\right)$

$-\dfrac{4}{7} \div \left(-\dfrac{11}{2}\right) = -\dfrac{4}{7} \cdot \left(-\dfrac{2}{11}\right) = \dfrac{8}{77}$

24. Find each quotient.

a. $\dfrac{15}{-3}$

$\dfrac{15}{-3} = -5$

b. $-\dfrac{30}{6}$

$-\dfrac{30}{6} = -5$

c. $\dfrac{-7.5}{-0.03}$

$\dfrac{-7.5}{-0.03} = 250$

d. $-\dfrac{1}{9} \div \left(-\dfrac{2}{3}\right)$

$-\dfrac{1}{9} \div \left(-\dfrac{2}{3}\right) = -\dfrac{1}{9} \cdot \left(-\dfrac{3}{2}\right) = \dfrac{1}{6}$

23. Find each quotient using the definition of division.

a. $\dfrac{30}{5}$

_____

b. $\dfrac{6(-9)}{9}$

_____

c. $\dfrac{-2.79}{-9}$

_____

d. $-\dfrac{6}{7} \div \left(-\dfrac{5}{6}\right)$

_____

24. Find each quotient.

a. $\dfrac{-18}{-6}$

_____

b. $\dfrac{16}{-8}$

_____

c. $\dfrac{-18.3}{-6.1}$

_____

d. $-\dfrac{2}{5} \div \left(-\dfrac{11}{10}\right)$

_____

CHAPTER 1 Equations and Inequalities

Name: _____ Date: _____
Instructor: _____ Section: _____

Practice Problems

For extra help for exercises 1–15, see the videos on properties of real numbers in your MyMathLab course.

Complete each statement. Use a commutative property.

1. $y + 4 =$ _____ $+ y$ 1. _____

2. $5(2) =$ _____ (5) 2. _____

3. $-4(4+z) =$ _____ (-4) 3. _____

Complete each statement. Use an associative property.

4. $4(ab) =$ _____ $\cdot b$ 4. _____

5. $[x + (-4)] + 3y = x +$ _____   5. _____

6. $4r + (3s + 14t) =$ _____ $+ 14t$ 6. _____

Use an identity property to complete each statement.

7. $4 + 0 =$ _____   7. _____

8. _____ $\cdot 1 = 12$ 8. _____

Name: Date:
Instructor: Section:

Use an identity property to simplify the expression.

9. $\dfrac{30}{35}$ 9. _____

Complete the statements so that they are examples of either an identity property or an inverse property. Identify which property is used.

10. $-4 + \underline{} = 0$ 10. _____

11. $-9 + \underline{} = -9$ 11. _____

12. $-\dfrac{3}{5} \cdot \underline{} = 1$ 12. _____

Use the distributive property to rewrite each expression. Simplify if possible.

13. $n(2a - 4b + 6c)$ 13. _____

14. $-2(5y - 9z)$ 14. _____

15. $-(-2k + 7)$ 15. _____

For extra help for exercises 16–18, see the videos on evaluating exponential expressions in your MyMathLab course.

Find the value of each exponential expression.

16. 3^3 16. _____

17. $\left(\dfrac{2}{3}\right)^4$ 17. _____

Copyright © 2015 Pearson Education, Inc.

CHAPTER 1 Equations and Inequalities

Name: Date:
Instructor: Section:

18. $(0.4)^2$

18. _____

For extra help for exercises 19–21, see the videos on order of operations in your MyMathLab course.

Perform the indicated operations.

19. $-4[(-2)(7)-2]$

19. _____

20. $\dfrac{-7(2)-(-3)}{5+(-3)}$

20. _____

21. $\dfrac{-4[8-(-3+7)]}{-6[3-(-2)]-3(-3)}$

21. _____

For extra help for exercises 22–24, see the videos on like terms in your MyMathLab course.

Identify each group of terms as **like** *or* **unlike**.

22. $4x^2, -7x^2$

22. _____

23. $-8m, -8m^2$

23. _____

24. $7xy, -6xy^2$

24. _____

For extra help for exercises 25–27, see the videos on simplifying expressions in your MyMathLab course.

Simplify.

25. $12y - 7y^2 + 4y - 3y^2$

25. _____

Copyright © 2015 Pearson Education, Inc.

Name: Date:
Instructor: Section:

26. $-4(x+4)+2(3x+1)$ 26. _____

27. $2.5(3y+1)-4.5(2y-3)$ 27. _____

For extra help for exercises 28–33, see the videos on adding real numbers in your MyMathLab course.

Use a number line to find the sum.

28. $-8+5$ 28. _____

Find each sum.

29. $-7+(-11)$ 29. _____

30. $-9+(-9)$ 30. _____

31. $-2\frac{3}{8}+\left(-3\frac{1}{4}\right)$ 31. _____

32. $\frac{7}{12}+\left(-\frac{3}{4}\right)$ 32. _____

33. $-\frac{4}{7}+\frac{3}{5}$ 33. _____

IRW-18   CHAPTER 1 Equations and Inequalities

Name: Date:
Instructor: Section:

For extra help for exercises 34–39, see the videos on subtracting real numbers in your MyMathLab course.

Use a number line to find the difference.

34. $8 - 5$ **34.**

35. $7 - 10$ **35.**

36. $-5 - 2$ **36.** _____

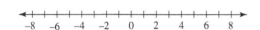

Find each difference.

37. $22 - (-24)$ **37.** _____

38. $-3.2 - (-7.6)$ **38.** _____

39. $3\frac{3}{4} - \left(-2\frac{1}{8}\right)$ **39.** _____

For extra help for exercises 40–45, see the videos on multiplying real numbers in your MyMathLab course.

Find each product.

40. $7(-4)$ **40.** _____

Name: Date:
Instructor: Section:

41. $\left(\dfrac{1}{5}\right)\left(-\dfrac{2}{3}\right)$ 41. _____

42. $(-3.2)(4.1)$ 42. _____

43. $(-4)(-10)$ 43. _____

44. $\left(-\dfrac{2}{7}\right)\left(-\dfrac{14}{5}\right)$ 44. _____

45. $(-0.4)(-3.4)$ 45. _____

For extra help for exercises 46–48, see the videos on dividing real numbers in your MyMathLab course.

Find each quotient.

46. $\dfrac{-120}{-20}$ 46. _____

47. $\dfrac{0}{-2}$ 47. _____

48. $\dfrac{10}{0}$ 48. _____

Chapter 1 Equations and Inequalities

1.2R Operations with Decimals; Translating Word Phrases into Algebraic Expressions and Equations; Formulas and Percent

Key Terms

Use the vocabulary terms listed below to complete each statement in exercises 1–10.

decimal places	factor	product
variable	constant	algebraic expression
equation	solution	percent equation
percent		

1. Each number in a multiplication problem is called a _____.

2. When multiplying decimal numbers, first multiply the numbers, then find the total number of _____ in both factors.

3. The answer to a multiplication problem is called the _____.

4. A(n) _____ is a statement that says two expressions are equal.

5. A _____ is a symbol, usually a letter, used to represent an unknown number.

6. A collection of numbers, variables, operation symbols, and grouping symbols is an_____.

7. Any value of a variable that makes an equation true is a(n) _____ of the equation.

8. A _____ is a fixed, unchanging number.

9. A number written with a _____ sign means "divided by 100".

10. The _____ is $part = percent \cdot whole$.

Name: Date:
Instructor: Section:

Guided Examples

Review these examples:

1. Find each sum.

   a. 29.73 and 56.84

   Step 1 Write the numbers in columns with the decimal points lined up.
   ```
     29.73
   + 56.84
   ```
   Step 2 Add as if these were whole numbers.
   Step 3 Line up decimal point in answer under the decimal points in problem.
   ```
     1 1
     29.73
   + 56.84
     86.57
   ```

   b. 8.437 + 5.361 + 13.295

   Write the numbers vertically with decimal points lined up. Then add.
   ```
     1 1  1 1
      8.437
      5.361
   + 13.295
     27.093
   ```

2. Find each sum.

   a. 6.7 + 0.41

   There are two decimal places in 0.41, so write a 0 in the hundredths place in 6.7 so it has two decimal places also.
   ```
     6.70
   + 0.41
     7.11
   ```

   b. 12 + 9.36 + 3.754

   Write in zeros so that all the addends have three decimal places.
   ```
     12.000
      9.360
   +  3.754
     25.114
   ```

Now Try:

1. Find each sum.

   a. 12.687 + 2.943

   _____

   b. 0.428 + 16.005 + 5.276

   _____

2. Find each sum.

   a. 7.53 + 29.314

   _____

   b. 0.631 + 999.3 + 14

   _____

CHAPTER 1 Equations and Inequalities

Name: Date:
Instructor: Section:

3. Find each difference.

  a. 14.32 from 36.74

Step 1 Line up decimal points.
```
  36.74
− 14.32
```
Step 2 Both numbers have two decimal places; no need to write in zeros.
Step 3 Line up decimal point in answer.
```
  36.74
− 14.32
  22.42
```

  b. 167.53 minus 69.85

Regrouping is needed here.
```
  0 15 16  14 13
  1̸ 6̸ 7̸ . 5̸ 3̸
−    6 9 . 8 5
     9 7 . 6 8
```
Line up decimal point in answer.

4. Find each difference.

  a. 19.6 from 38.264

Line up decimal points and write in zeros so both numbers have three decimal places.
```
  38.264
− 19.600
  18.664
```

  b. 28.8 − 19.963

Write in two zeros and subtract as usual.
```
  28.800
− 19.963
   8.837
```

  c. 16 less 7.54

Write a decimal point and two zeros after 16. Subtract as usual.
```
  16.00
−  7.54
   8.46
```

3. Find each difference.

  a. 7.352 from 18.964

  _____

  b. 50.43 minus 39.86

  _____

4. Find each difference.

  a. 3.87 from 8.524

  _____

  b. 20 − 16.74

  _____

  c. 1 less 0.499

  _____

Copyright © 2015 Pearson Education, Inc.

Name: Date:
Instructor: Section:

5. Find the product of 6.23 and 5.4.

   Step 1 Multiply the numbers as if they were whole numbers.

   $$\begin{array}{r} 6.23 \\ \times\ \ \ 5.4 \\ \hline 2492 \\ 3115\ \ \\ \hline 33642 \end{array}$$

   Step 2 Count the total number of decimal places in both factors.

   $$\begin{array}{r} 6.23 \\ \times\ \ \ 5.4 \\ \hline 2492 \\ 3115\ \ \\ \hline 33642 \end{array}$$ ← 2 decimal places
   ← 1 decimal place
   3 total decimal places

   Step 3 Count over 3 places in the product and write the decimal point. Count from right to left.

   $$\begin{array}{r} 6.23 \\ \times\ \ \ 5.4 \\ \hline 2492 \\ 3115\ \ \\ \hline 33.642 \end{array}$$ ← 2 decimal places
   ← 1 decimal place
   3 total decimal places

5. Find the product of 2.51 and 4.3.

   _____

6. Find the product: $(0.035)(0.07)$.

   Start by multiply, then count decimal places.

   $$\begin{array}{r} 0.035 \\ \times\ \ 0.07 \\ \hline 245 \end{array}$$ ← 3 decimal places
   ← 2 decimal places
   5 total decimal places

   After multiplying, the answer has only three decimal places, but five are needed, so write two zeros on the left side of the answer. Then count over 5 places and write in the decimal point.

   $$\begin{array}{r} 0.035 \\ \times\ \ 0.07 \\ \hline .00245 \end{array}$$ ← 3 decimal places
   ← 2 decimal places
   5 total decimal places

   The final product is 0.00245, which has five decimal places.

6. Find the product $(0.062)(0.03)$.

   _____

CHAPTER 1 Equations and Inequalities

Name: Date:
Instructor: Section:

7. Find each quotient. Check the quotients by multiplying.

   a. 24.48 by 4

   Rewrite the division problem.
   Step 1 Write the decimal point in the quotient directly above the decimal point in the dividend.

   $$4\overline{)24.48}$$ (with decimal point above)

   Step 2 Divide as if the numbers were whole numbers.

   $$\begin{array}{r}6.12\\4\overline{)24.48}\end{array}$$

   Check by multiplying the quotient times the divisor.
   $$\begin{array}{r}6.12\\\times\ \ 4\\\hline 24.48\end{array}$$
   The quotient (answer) is 6.12.

   b. $8\overline{)425.6}$

   Write the decimal point in the quotient above the decimal point in the dividend. Then divide as if the number were whole numbers.

   $$\begin{array}{r}53.2\\8\overline{)425.6}\\\underline{40}\\25\\\underline{24}\\16\\\underline{16}\\0\end{array}$$ Check: $\begin{array}{r}53.2\\\times\ \ 8\\\hline 425.6\end{array}$

   The quotient is 53.2.

7. Find each quotient. Check the quotients by multiplying.

   a. 9.891 by 7

   _____

   b. $5\overline{)75.15}$

   _____

Name: Date:
Instructor: Section:

8. Divide 1.35 by 4. Check the quotient by multiplying.

Divide.

```
    0.33
4)1.35
    1 2
    ─────
      15
      12
      ──
       3
```

Write a 0 after the 5 in the dividend so you can continue dividing. Keep writing more zeros in the dividend, if needed.

```
    0.3375              0.3375
4)1.3500    Check:    ×    4
    1 2               ────────
    ──                1.3500
     15
     12
     ──
      30
      28
      ──
       20
       20
       ──
        0
```

The quotient is 0.3375.

8. Divide 1008.9 by 50. Check the quotient by multiplying.

_____

CHAPTER 1 Equations and Inequalities

Name: Date:
Instructor: Section:

8. Divide 8.87 by 9. Round the quotient to the nearest thousandth.

Write extra zeros in the dividend so you can continue dividing.

$$\begin{array}{r} 0.9855 \\ 9\overline{)8.8700} \\ \underline{8\ 1} \\ 77 \\ \underline{72} \\ 50 \\ \underline{45} \\ 50 \\ \underline{45} \\ 5 \end{array}$$

Notice that the digit 5 in the answer is repeating. There are two ways to show that the answer is a repeating decimal that goes on forever.

  0.9855... or $0.98\overline{5}$

To round to thousandths, divide out one more place, to ten-thousandths.

  $8.87 \div 9 = 0.9855...$ rounds to 0.986.

Check the answer by multiplying 0.986 by 9. Because 0.986 is a rounded answer, the check will not give exactly 8.87, but it should be very close.

  $(0.986)(9) = 8.874$

10. a. $0.005\overline{)41.2}$

Move the decimal point in the divisor three places to the right so 0.005 becomes the whole number 5. Move the decimal point in the dividend the same number of places and write in two extra 0s.

$$\begin{array}{r} 8240. \\ 5\overline{)41200.} \end{array}$$

9. Divide 302.24 by 18. Round the quotient to the nearest thousandth.

  _____

10. a. $0.0024\overline{)48.984}$

  _____

Name: Date:
Instructor: Section:

b. Divide 7 by 2.4. Round to the nearest hundredth.

Move the decimal point in the divisor one place to the right so 2.4 becomes the whole number 24. The decimal point in the dividend starts on the right side of 7 and is also moved one place to the right.
(Remember, in order to round to hundredths, divide out one more place, to thousandths.)

$$\begin{array}{r} 2.916 \\ 24\overline{)70.000} \\ \underline{48} \\ 220 \\ \underline{216} \\ 40 \\ \underline{24} \\ 160 \\ \underline{144} \\ 16 \end{array}$$

Round the quotient. It is 2.92.

11. Write each word phrase as an algebraic expression, using x as the variable.

  a. The sum of a number and 7

  $x + 7$, or $7 + x$

  b. 8 minus a number

  $8 - x$

  c. A number subtracted from 13

  $13 - x$

  d. The product of 15 and a number

  $15 \cdot x$, or $15x$

  e. 9 divided by a number

  $9 \div x$, or $\dfrac{9}{x}$

b. Divide 8 by 4.5. Round to the nearest hundredth.

_____

11. Write each word phrase as an algebraic expression, using x as the variable.

  a. The sum of a number and 4

  _____

  b. 10 minus a number

  _____

  c. A number subtracted from 12

  _____

  d. The product of 20 and a number

  _____

  e. 10 divided by a number

  _____

Copyright © 2015 Pearson Education, Inc.

CHAPTER 1 Equations and Inequalities

Name: Date:
Instructor: Section:

f. The product of 3 and the difference between a number and 5

$3 \cdot (x-5)$, or $3(x-5)$

f. The product of 5 and the difference between a number and 6

_____

12. Write each word sentence as an equation. Use x as the variable.

a. Four times the sum of a number and five is eleven.

Four times   the sum of a number and five   is   eleven.
   ↓   ↓   ↓   ↓
   4 ·   ($x+5$)   =   11

$4(x+5) = 11$

b. Eight more than six times a number is 90.

Start with $6x$ and then add 8 to it.
$6x + 8 = 90$

c. Nine less than five times a number is equal to twenty.

Five times a number less nine is equal to twenty.
     ↓   ↓   ↓   ↓   ↓
    $5x$ –   9   =   20

$5x - 9 = 20$

13. a. Find 16% of $2300.

Write 16% as the decimal 0.16. The whole, which comes after the word of, is 2300. Next, use the percent equation. Let x represent the unknown part.

   part = percent × whole
   $x = (0.16)(2300)$

Multiply 0.16 and 2300.
   $x = 368$
16% of $2300 is $368.

12. Write each word sentence as an equation. Use x as the variable.

a. Three times the sum of a number and seven is thirty.

_____

b. Three more than five times a number is 50.

_____

c. Seven less than three times a number is equal to twelve.

_____

13. a. Find 18% of $350.

_____

Name: Date:
Instructor: Section:

b. Find 120% of 90 cartons.

Write 120% as the decimal 1.20. The whole is 90. Let x represent the unknown part.
$$\text{part} = \text{percent} \times \text{whole}$$
$$x = (1.20)(90)$$
$$x = 108$$
120% of 90 cartons is 108 cartons.

b. Find 130% of 70 packages.

_____

c. Find 0.8% of 3500 cases.

Write 0.8% as the decimal 0.008. The whole is 3500. Let x represent the unknown part.
$$\text{part} = \text{percent} \times \text{whole}$$
$$x = (0.008)(3500)$$
$$x = 28$$
0.8% of 3500 cases is 28 cases.

c. Find 0.6% of 31,000 students.

_____

14. a. 160 gallons is 25% of what number of gallons?

The part is 160 and the percent is 25% or the decimal 0.25. The whole is unknown.
   160 is 25% of what number?
Now, use the percent equation.
$$\text{part} = \text{percent} \times \text{whole}$$
$$160 = (0.25)(x)$$
$$\frac{160}{0.25} = \frac{\cancel{(0.25)}^{1}(x)}{\cancel{0.25}_{1}}$$
$$640 = x$$
160 gallons is 25% of 640 gallons.

14. a. 45 units is 37.5% of what number of units.

_____

Copyright © 2015 Pearson Education, Inc.

b. 25 employees is 5% of what number of employees.

Write 5% as 0.05. The part is 25. Use the percent equation to find the whole.

part = percent × whole

$25 = (0.05)(x)$

$$\frac{25}{0.05} = \frac{(0.05)(x)}{0.05}$$

$500 = x$

25 employees is 5% of 500 employees.

c. 75 is $6\frac{1}{4}\%$ of what number?

Write $6\frac{1}{4}\%$ as 6.25%, or the decimal 0.0625.

The part is 75. Use the percent equation.

part = percent × whole

$75 = (0.0625)(x)$

$$\frac{75}{0.0625} = \frac{(0.0625)(x)}{0.0625}$$

$1200 = x$

75 is $6\frac{1}{4}\%$ of 1200.

15. a. $700 is what percent of $2800?

Because $2800 follows of, the whole is $2800. The part is $700, and the percent is unknown. Use the percent formula.

part = percent · whole

$700 = x \cdot 2800$

$$\frac{700}{2800} = \frac{x \cdot 2800}{2800}$$

$0.25 = x$

$700 is 25% of $2800.

---

b. 330 points is 13.25% of how many points

_____

c. 1750% of what number is 1050?

_____

15. a. 145 miles is what percent of 500 miles?

_____

Name: Date:
Instructor: Section:

b. What percent of 250 is 115?

The whole is 250 and the part is 115. Let x represent the unknown percent.
$$\text{part} = \text{percent} \cdot \text{whole}$$
$$115 = x \cdot 250$$
$$\frac{115}{250} = \frac{x \cdot \cancel{250}}{\cancel{250}}$$
$$0.46 = x$$
46% of 250 is 115.

c. What percent of 1500 calories is 1800 calories?

The whole is 1500 and the part is 1800. Let x represent the unknown percent.
$$\text{part} = \text{percent} \cdot \text{whole}$$
$$1800 = x \cdot 1500$$
$$\frac{1800}{1500} = \frac{x \cdot \cancel{1500}}{\cancel{1500}}$$
$$1.2 = x$$
1.2 is 120%
120% of 1500 calories is 1800 calories.

d. $2400 is what percent of $35,000?

Since $35,000 follows of, the whole is $35,000. The part is $2400.
$$\text{part} = \text{percent} \cdot \text{whole}$$
$$2400 = x \cdot 35,000$$
$$\frac{2400}{35,000} = \frac{x \cdot \cancel{35,000}}{\cancel{35,000}}$$
$$0.07 = x$$
7% of $35,000 is $2450.

b. What percent of $225,000 is $45,000?

_____

c. What percent of 90 is 288?

_____

d. 24 miles is what percent of 6000 miles?

_____

CHAPTER 1 Equations and Inequalities

Name: Date:
Instructor: Section:

Practice Problems

For extra help for exercises 1–3, see the videos on adding decimals in your MyMathLab course.

Find each sum.

1. $43.96 + 48.53$ 1. _____

2. $87.6 + 90.4$ 2. _____

3. $45.83 + 20.923 + 5.7$ 3. _____

For extra help for exercises 4–6, see the videos on subtracting decimals in your MyMathLab course.

Find each difference.

4. $84.6 - 18.1$ 4. _____

5. $69.524 - 26.958$ 5. _____

6. $71 - 12.68$ 6. _____

For extra help for exercises 7–9, see the videos on multiplying decimals in your MyMathLab course.

Find each product.

7. $\begin{array}{r} 19.3 \\ \times\ 4.7 \end{array}$ 7. _____

Copyright © 2015 Pearson Education, Inc.

Name: Date:
Instructor: Section:

8. 0.682
   × 3.9

8. _____

9. (0.074)(0.05)

9. _____

For extra help for exercises 10–15, see the videos on dividing decimals in your MyMathLab course.

Find each quotient. Round answers to the nearest thousandth, if necessary.

10. $5\overline{)34.8}$

10. _____

11. $11\overline{)46.98}$

11. _____

12. $33\overline{)77.847}$

12. _____

Find each quotient. Round answers to the nearest thousandth, if necessary.

13. $0.9\overline{)3.4166}$

13. _____

14. $3.4\overline{)436.05}$

14. _____

Copyright © 2015 Pearson Education, Inc.

IRW-34 CHAPTER 1 Equations and Inequalities

Name: Date:
Instructor: Section:

15. $0.07 \div 0.00043$ 15. _____

For extra help for exercises 16–21, see the videos on writing word phrases or sentences as algebraic expressions or equations in your MyMathLab course.

Write each word phrase as an algebraic expression. Use x as the variable.

16. Ten times a number, added to 21   16. _____

17. 11 fewer than eight times a number   17. _____

18. Half a number subtracted from two-thirds of the number   18. _____

Write each word sentence as an equation. Use x as the variable.

19. Ten divided by a number is two more than the number.   19. _____

20. The product of six and five more than a number is nineteen.   20. _____

21. Seven times a number subtracted from 61 is 13 plus the number.   21. _____

For extra help for exercises 22–30, see the videos on using the percent equation in your MyMathLab course.

Find the part using the percent equation. Round to the nearest tenth, if necessary.

22. 9% of 240   22. _____

23. 140% of 76   23. _____

Copyright © 2015 Pearson Education, Inc.

Name: Date:
Instructor: Section:

24. 0.4% of 350   **24.** _____

Find the whole using the percent equation. Round to the nearest tenth, if necessary.

25. 64 is 40% of what number?   **25.** _____

26. 75% of what number is 1125?   **26.** _____

27. 35 is 153% of what number?   **27.** _____

Find the percent using the percent equation. Round to the nearest tenth, if necessary.

28. 15 is what percent of 75?   **28.** _____

29. What percent of 160 is 8?   **29.** _____

30. What percent of 18 is 44?   **30.** _____

Chapter 1 Equations and Inequalities

1.3R Radical Notation; Simplifying Square Root Radicals; Rationalizing Square Root Denominators; Addition, Subtraction, and Multiplication of Square Root Radicals; Product Rule for Exponents; Zero Exponent Rule; Negative Exponent Rule; Add, Subtract, and Multiply Polynomials

Key Terms

Use the vocabulary terms listed below to complete each statement in exercises 1–23.

square root	principal square root	radicand
radical	radical expression	perfect square
irrational number	rationalizing the denominator	
like radicals	unlike radicals	product rule for exponents
term	like terms	polynomial
descending powers		degree of a term
degree of a polynomial		monomial
binomial	trinomial	FOIL
outer product	inner product	

1. The number or expression inside a radical sign is called the _____.

2. A number with a rational square root is called a _____.

3. The number b is a _____ of a if $b^2 = a$.

4. The expression $\sqrt[n]{a}$ is called a _____.

5. The positive square root of a number is its _____.

6. A real number that is not rational is called an _____.

7. A _____ is a radical sign and the number or expression in it.

8. The process of removing radicals from the denominator so that the denominator contains only rational quantities is called _____.

9. The expressions $2\sqrt{2}$ and $6\sqrt[3]{2}$ are _____.

10. The expressions $2\sqrt{2}$ and $7\sqrt{2}$ are _____.

Name: Date:
Instructor: Section:

11. The statement "If m and n are any integers, then $a^m \cdot a^n = a^{m+n}$ is an example of the _____.

12. The _____ is the sum of the exponents on the variables in that term.

13. A polynomial in x is written in _____ if the exponents on x in its terms are decreasing order.

14. A _____ is a number, a variable, or a product or quotient of a number and one or more variables raised to powers.

15. A polynomial with exactly three terms is called a _____.

16. A _____ is a term, or the sum of a finite number of terms with whole number exponents.

17. A polynomial with exactly one term is called a _____.

18. The _____ is the greatest degree of any term of the polynomial.

19. A _____ is a polynomial with exactly two terms.

20. Terms with exactly the same variables (including the same exponents) are called _____.

21. The _____ of $(2y-5)(y+8)$ is $-5y$.

22. _____ is a shortcut method for finding the product of two binomials.

23. The _____ of $(2y-5)(y+8)$ is $16y$.

Guided Examples

Review these examples:

1. Find the square roots of 64.

    What number multiplied by itself equals 64?
    $8^2 = 64$ and $(-8)^2 = 64$.
    Thus, 64 has two square roots: 8 and −8.

Now Try:

1. Find the square roots of 81.

    _____

CHAPTER 1 Equations and Inequalities

Name: Date:
Instructor: Section:

1. Find each square root.

   a. $\sqrt{121}$

   $11^2 = 121$, so $\sqrt{121} = 11$.

   b. $-\sqrt{576}$

   This is the negative square root of 576. Since $\sqrt{576} = 24$, then $-\sqrt{576} = -24$.

   c. $\sqrt{\dfrac{16}{25}}$

   $\sqrt{\dfrac{16}{25}} = \dfrac{4}{5}$

   d. $\sqrt{0.36}$

   $\sqrt{0.36} = 0.6$

2. Find the square of each radical expression.

   a. $\sqrt{17}$

   The square of $\sqrt{17}$ is $\left(\sqrt{17}\right)^2 = 17$.

   b. $-\sqrt{31}$

   $\left(-\sqrt{31}\right)^2 = 31$

   c. $\sqrt{w^2 + 3}$

   $\left(\sqrt{w^2+3}\right)^2 = w^2 + 3$

3. Tell whether each square root is rational, irrational, or not a real number.

   a. $\sqrt{5}$

   Because 5 is not a perfect square, $\sqrt{5}$ is irrational.

---

2. Find each square root.

   a. $\sqrt{169}$

   _____

   b. $\sqrt{1681}$

   _____

   c. $\sqrt{\dfrac{9}{49}}$

   _____

   d. $\sqrt{0.64}$

   _____

3. Find the square of each radical expression.

   a. $\sqrt{19}$

   _____

   b. $-\sqrt{37}$

   _____

   c. $\sqrt{n^2 + 5}$

   _____

4. Tell whether each square root is rational, irrational, or not a real number.

   a. $\sqrt{11}$

   _____

Copyright © 2015 Pearson Education, Inc.

Name: Date:
Instructor: Section:

b. $\sqrt{81}$

81 is a perfect square, 9^2, so $\sqrt{81} = 9$ is a rational number.

c. $\sqrt{-16}$

There is no real number whose square is -16. Therefore, $\sqrt{-16}$ is not a real number.

5. Find each square root. In parts (c) and (d), m is a real number.

a. $\sqrt{33^2}$

$\sqrt{33^2} = |33| = 33$

b. $\sqrt{(-33)^2}$

$\sqrt{(-33)^2} = |-33| = 33$

c. $\sqrt{m^2}$

$\sqrt{m^2} = |m|$

d. $\sqrt{(-m)^2}$

$\sqrt{(-m)^2} = |-m| = |m|$

6. Simplify.

a. $\sqrt{90}$

$\sqrt{90} = \sqrt{9 \cdot 10}$
$= \sqrt{9} \cdot \sqrt{10}$
$= 3\sqrt{10}$

b. $\sqrt{288}$

$\sqrt{288} = \sqrt{144 \cdot 2}$
$= \sqrt{144} \cdot \sqrt{2}$
$= 12\sqrt{2}$

b. $\sqrt{100}$

c. $\sqrt{-9}$

5. Find each square root. In parts (c) and (d), n is a real number.

a. $\sqrt{73^2}$

b. $\sqrt{(-37)^2}$

c. $\sqrt{n^2}$

d. $\sqrt{(-n)^2}$

6. Simplify.

a. $\sqrt{84}$

b. $\sqrt{162}$

Copyright © 2015 Pearson Education, Inc.

CHAPTER 1 Equations and Inequalities

Name: Date:
Instructor: Section:

   c. $\sqrt{35}$

No perfect square (other than 1) divides into 35, so $\sqrt{35}$ cannot be simplified further.

7. Simplify. Assume that all variables represent positive real numbers.

   a. $\sqrt{81x^3}$

$\sqrt{81x^3} = \sqrt{9^2 \cdot x^2 \cdot x} = 9x\sqrt{x}$

   b. $\sqrt{56x^7 y^6}$

$\sqrt{56x^7 y^6} = \sqrt{4 \cdot 14 \cdot (x^3)^2 \cdot x \cdot (y^3)^2}$
$= 2x^3 y^3 \sqrt{14x}$

8. Rationalize each denominator.

   a. $\dfrac{4}{\sqrt{3}}$

$\dfrac{4}{\sqrt{3}} = \dfrac{4 \cdot \sqrt{3}}{\sqrt{3} \cdot \sqrt{3}} = \dfrac{4\sqrt{3}}{3}$

   b. $\dfrac{\sqrt{3}}{8\sqrt{5}}$

$\dfrac{\sqrt{3}}{8\sqrt{5}} = \dfrac{\sqrt{3} \cdot \sqrt{5}}{8\sqrt{5} \cdot \sqrt{5}} = \dfrac{\sqrt{15}}{8 \cdot 5} = \dfrac{\sqrt{15}}{40}$

   c. $-\dfrac{15}{\sqrt{27}}$

First, simplify the denominator.
$-\dfrac{15}{\sqrt{27}} = -\dfrac{15}{\sqrt{9 \cdot 3}} = -\dfrac{15}{3\sqrt{3}} = -\dfrac{5}{\sqrt{3}}$

Now, rationalize the denominator.
$-\dfrac{5}{\sqrt{3}} = -\dfrac{5 \cdot \sqrt{3}}{\sqrt{3} \cdot \sqrt{3}} = -\dfrac{5\sqrt{3}}{3}$

---

   c. $\sqrt{95}$

   _____

7. Simplify. Assume that all variables represent positive real numbers.

   a. $\sqrt{100 y^3}$

   _____

   b. $\sqrt{48 m^5 r^9}$

   _____

8. Rationalize each denominator.

   a. $\dfrac{2}{\sqrt{15}}$

   _____

   b. $\dfrac{6\sqrt{3}}{\sqrt{14}}$

   _____

   c. $-\dfrac{3}{\sqrt{75}}$

   _____

Copyright © 2015 Pearson Education, Inc.

Name: Date:
Instructor: Section:

9. Simplify each radical.

a. $-\sqrt{\dfrac{27}{98}}$

$-\sqrt{\dfrac{27}{98}} = -\dfrac{\sqrt{27}}{\sqrt{98}}$

$= -\dfrac{\sqrt{9 \cdot 3}}{\sqrt{49 \cdot 2}}$

$= -\dfrac{3\sqrt{3}}{7\sqrt{2}}$

$= -\dfrac{3\sqrt{3} \cdot \sqrt{2}}{7\sqrt{2} \cdot \sqrt{2}}$

$= -\dfrac{3\sqrt{6}}{7 \cdot 2}$

$= -\dfrac{3\sqrt{6}}{14}$

b. $\sqrt{\dfrac{20b^4}{a^5}}$, $a > 0$, $b \geq 0$

$\sqrt{\dfrac{20b^4}{a^5}} = \dfrac{\sqrt{20b^4}}{\sqrt{a^5}}$

$= \dfrac{\sqrt{4b^4 \cdot 5}}{\sqrt{a^4 \cdot a}}$

$= \dfrac{2b^2\sqrt{5}}{a^2\sqrt{a}}$

$= \dfrac{2b^2\sqrt{5} \cdot \sqrt{a}}{a^2\sqrt{a} \cdot \sqrt{a}}$

$= \dfrac{2b^2\sqrt{5a}}{a^3}$

9. Simplify each radical.

a. $-\sqrt{\dfrac{45}{32}}$

_____

b. $\sqrt{\dfrac{162x^3}{t^5}}$, $x \geq 0$, $t > 0$

_____

Copyright © 2015 Pearson Education, Inc.

CHAPTER 1 Equations and Inequalities

Name: Date:
Instructor: Section:

10. Add or subtract to simplify each radical expression. Assume that all variables represent positive real numbers.

   a. $3\sqrt{13}+5\sqrt{52}$

   $3\sqrt{13}+5\sqrt{52} = 3\sqrt{13}+5\sqrt{4}\sqrt{13}$
   $= 3\sqrt{13}+5\cdot 2\sqrt{13}$
   $= 3\sqrt{13}+10\sqrt{13}$
   $= (3+10)\sqrt{13}$
   $= 13\sqrt{13}$

   b. $\sqrt{48x}-\sqrt{12x}, \ x \geq 0$

   $\sqrt{48x}-\sqrt{12x} = \sqrt{16}\cdot\sqrt{3x}-\sqrt{4}\cdot\sqrt{3x}$
   $= 4\sqrt{3x}-2\sqrt{3x}$
   $= (4-2)\sqrt{3x}$
   $= 2\sqrt{3x}$

   c. $7\sqrt{3}-6\sqrt{21}$

   The radicands differ and are already simplified, so this expression cannot be simplified further.

   d. $\dfrac{\sqrt{32}}{3}+\dfrac{\sqrt{8}}{\sqrt{18}}$

   $\dfrac{\sqrt{32}}{3}+\dfrac{\sqrt{8}}{\sqrt{18}} = \dfrac{\sqrt{16\cdot 2}}{3}+\dfrac{\sqrt{4\cdot 2}}{\sqrt{9\cdot 2}}$
   $= \dfrac{4\sqrt{2}}{3}+\dfrac{2\sqrt{2}}{3\sqrt{2}}$
   $= \dfrac{4\sqrt{2}}{3}+\dfrac{2}{3}$
   $= \dfrac{4\sqrt{2}+2}{3}$

10. Add or subtract to simplify each radical expression. Assume that all variables represent positive real numbers.

   a. $3\sqrt{54}-5\sqrt{24}$

   _____

   b. $3\sqrt{18z}+2\sqrt{8z}, \ z \geq 0$

   _____

   c. $3\sqrt{7}+2\sqrt{6}$

   _____

   d. $\sqrt{\dfrac{10}{18}}+\dfrac{\sqrt{15}}{\sqrt{27}}$

   _____

Name: Date:
Instructor: Section:

11. Multiply, using the FOIL method.

a. $\left(2\sqrt{3}-\sqrt{7}\right)\left(\sqrt{5}+\sqrt{2}\right)$

$\left(2\sqrt{3}-\sqrt{7}\right)\left(\sqrt{5}+\sqrt{2}\right)$
$= 2\sqrt{3}\cdot\sqrt{5}+2\sqrt{3}\cdot\sqrt{2}-\sqrt{7}\cdot\sqrt{5}-\sqrt{7}\sqrt{2}$
$= 2\sqrt{15}+2\sqrt{6}-\sqrt{35}-\sqrt{14}$

b. $\left(5-\sqrt{3}\right)\left(\sqrt{2}+\sqrt{5}\right)$

$\left(5-\sqrt{3}\right)\left(\sqrt{2}+\sqrt{5}\right)$
$= 5\cdot\sqrt{2}+5\cdot\sqrt{5}-\sqrt{3}\cdot\sqrt{2}-\sqrt{3}\cdot\sqrt{5}$
$= 5\sqrt{2}+5\sqrt{5}-\sqrt{6}-\sqrt{15}$

c. $\left(\sqrt{6}+\sqrt{5}\right)\left(\sqrt{6}-\sqrt{5}\right)$

This is the difference of squares.
$\left(\sqrt{6}+\sqrt{5}\right)\left(\sqrt{6}-\sqrt{5}\right)=\left(\sqrt{6}\right)^2-\left(\sqrt{5}\right)^2$
$= 6-5$
$= 1$

d. $\left(\sqrt{11}-6\right)^2$

$\left(\sqrt{11}-6\right)^2 = \left(\sqrt{11}-6\right)\left(\sqrt{11}-6\right)$
$= \sqrt{11}\cdot\sqrt{11}-6\sqrt{11}-6\sqrt{11}+6\cdot 6$
$= 11-12\sqrt{11}+36$
$= 47-12\sqrt{11}$

e. $\left(2+\sqrt[3]{5}\right)\left(2-\sqrt[3]{5}\right)$

$\left(2+\sqrt[3]{5}\right)\left(2-\sqrt[3]{5}\right)$
$= 2\cdot 2-2\sqrt[3]{5}+2\sqrt[3]{5}-\sqrt[3]{5}\cdot\sqrt[3]{5}$
$= 4-\sqrt[3]{5^2}$
$= 4-\sqrt[3]{25}$

11. Multiply, using the FOIL method.

a. $\left(\sqrt{5}+\sqrt{3}\right)\left(\sqrt{2}-\sqrt{11}\right)$

_____

b. $\left(3-\sqrt{2}\right)\left(2+\sqrt{7}\right)$

_____

c. $\left(\sqrt{14}-\sqrt{2}\right)\left(\sqrt{14}+\sqrt{2}\right)$

_____

d. $\left(3-\sqrt{2}\right)^2$

_____

e. $\left(4-\sqrt[3]{2}\right)\left(4+\sqrt[3]{2}\right)$

_____

Name: **Date:**
Instructor: **Section:**

f. $\left(\sqrt{m}-\sqrt{n}\right)\left(\sqrt{m}+\sqrt{n}\right)$

$\left(\sqrt{m}-\sqrt{n}\right)\left(\sqrt{m}+\sqrt{n}\right)$
$=\left(\sqrt{m}\right)^2-\left(\sqrt{n}\right)^2$
$=m-n, \quad m\geq 0 \text{ and } n\geq 0$

f. $\left(\sqrt{p}+\sqrt{q}\right)\left(\sqrt{p}-\sqrt{q}\right)$

_____

12. Use the product rule for exponents to simplify, if possible.

a. $8^4 \cdot 8^5$

$8^4 \cdot 8^5 = 8^{4+5}$
$ = 8^9$

b. $(-5)^8 (-5)^3$

$(-5)^8 (-5)^3 = (-5)^{8+3}$
$ = (-5)^{11}$

c. $x^3 \cdot x$

$x^3 \cdot x = x^3 \cdot x^1$
$ = x^{3+1}$
$ = x^4$

d. $m^7 m^8 m^9$

$m^7 m^8 m^9 = m^{7+8+9}$
$ = m^{24}$

e. $5^2 \cdot 4^3$

$5^2 \cdot 4^3 = 25 \cdot 64$
$ = 1600$

f. $5^2 + 5^3$

$5^2 + 5^3 = 25 + 125$
$ = 150$

12. Use the product rule for exponents to simplify, if possible.

a. $9^6 \cdot 9^7$

_____

b. $(-6)^2 (-6)^4$

_____

c. $x^5 \cdot x$

_____

d. $m^{11} m^9 m^7$

_____

e. $5^3 \cdot 2^2$

_____

f. $3^4 + 3^3$

_____

Name: Date:
Instructor: Section:

13. Evaluate.

    a. $75^0 = 1$

    b. $(-75)^0 = 1$

    c. $-75^0 = -(1)$ or -1

    d. $x^0 = 1$ $(x \neq 0)$

    e. $9x^0 = 9(1)$, or 9 $(x \neq 0)$

    f. $(9x)^0 = 1$ $(x \neq 0)$

14. Simplify by writing with positive exponents. Assume that all variables represent nonzero real numbers.

    a. 5^{-2}

    $5^{-2} = \dfrac{1}{5^2}$, or $\dfrac{1}{25}$

    b. 4^{-3}

    $4^{-3} = \dfrac{1}{4^3}$, or $\dfrac{1}{64}$

    c. $\left(\dfrac{1}{3}\right)^{-3}$

    $\left(\dfrac{1}{3}\right)^{-3} = 3^3$, or 27

13. Evaluate.

    a. 88^0

    b. $(-88)^0$

    c. -88^0

    d. a^0 $(a \neq 0)$

    e. $88a^0$ $(a \neq 0)$

    f. $(88a)^0$ $(a \neq 0)$

14. Simplify by writing with positive exponents. Assume that all variables represent nonzero real numbers.

    a. 8^{-2}

    b. 3^{-3}

    c. $\left(\dfrac{1}{5}\right)^{-2}$

Name: Date:
Instructor: Section:

d. $\left(\dfrac{3}{5}\right)^{-4}$

$\left(\dfrac{3}{5}\right)^{-4} = \left(\dfrac{5}{3}\right)^{4}$ The reciprocal of $\dfrac{3}{5}$ is $\dfrac{5}{3}$.

$= \dfrac{5^4}{3^4}$

$= \dfrac{625}{81}$

e. $\left(\dfrac{2}{3}\right)^{-5}$

$\left(\dfrac{2}{3}\right)^{-5} = \left(\dfrac{3}{2}\right)^{5}$

$= \dfrac{3^5}{2^5}$

$= \dfrac{243}{32}$

f. $5^{-1} - 3^{-1}$

$5^{-1} - 3^{-1} = \dfrac{1}{5} - \dfrac{1}{3}$

$= \dfrac{3}{15} - \dfrac{5}{15}$

$= -\dfrac{2}{15}$

g. $4q^{-3}$

$4q^{-3} = \dfrac{4}{1} \cdot \dfrac{1}{q^3}$

$= \dfrac{4}{q^3}$

d. $\left(\dfrac{6}{7}\right)^{-2}$

_____

e. $\left(\dfrac{3}{2}\right)^{-3}$

_____

f. $4^{-1} - 8^{-1}$

_____

g. $11p^{-4}$

_____

Name: Date:
Instructor: Section:

h. $\dfrac{1}{x^{-6}}$

$$\dfrac{1}{x^{-6}} = \dfrac{1^{-6}}{x^{-6}}$$
$$= \left(\dfrac{1}{x}\right)^{-6}$$
$$= x^6$$

i. $x^5 y^{-8}$

$$x^5 y^{-8} = \dfrac{x^5}{1} \cdot \dfrac{1}{y^8}$$
$$= \dfrac{x^5}{y^8}$$

15. Simplify. Assume that all variables represent nonzero real numbers.

   a. $\dfrac{3^{-4}}{7^{-2}} = \dfrac{7^2}{3^4}$, or $\dfrac{49}{81}$

   b. $\dfrac{a^{-6}}{b^{-1}} = \dfrac{b^1}{a^6}$, or $\dfrac{b}{a^6}$

   c. $\dfrac{x^{-3} y}{4 z^{-4}} = \dfrac{y z^4}{4 x^3}$

h. $\dfrac{1}{x^{-8}}$

_____

i. $a^{-5} b^4$

_____

15. Simplify. Assume that all variables represent nonzero real numbers.

   a. $\dfrac{6^{-2}}{5^{-3}}$

_____

   b. $\dfrac{x^{-7}}{y^{-1}}$

_____

   c. $\dfrac{p^{-3} q}{4 r^{-5}}$

_____

Copyright © 2015 Pearson Education, Inc.

CHAPTER 1 Equations and Inequalities

Name: Date:
Instructor: Section:

d. $\left(\dfrac{p}{3q}\right)^{-3}$

$\left(\dfrac{p}{3q}\right)^{-3} = \left(\dfrac{3q}{p}\right)^{3}$

$= \dfrac{3^3 q^3}{p^3}$

$= \dfrac{27 q^3}{p^3}$

d. $\left(\dfrac{a}{5b}\right)^{-4}$

_____

16. Add vertically.

  a. $5x^4 - 7x^3 + 9$ and $-3x^4 + 8x^3 - 7$

Write like terms in columns.

$\quad 5x^4 - 7x^3 + 9$
$\underline{-3x^4 + 8x^3 - 7}$

Now add, column by column.

$\quad 5x^4 \quad -7x^3 \quad\;\; 9$
$\underline{-3x^4 \quad\;\; 8x^3 \quad -7}$
$\quad 2x^4 \qquad x^3 \qquad 2$

Add the three sums together to obtain the answer.

$2x^4 + x^3 + 2$

  b. $3x^3 + 7x + 5$ and $x^4 - 6x$

Write like terms and add column by column.

$\quad\quad 3x^3 \;\; +7x \;\; +5$
$\underline{x^4 \quad\quad\quad\; -6x \quad\quad}$
$x^4 + 3x^3 \;\; +x \;\; +5$

16. Add vertically.

  a. $8x^3 - 9x^2 + x$ and
$-3x^3 + 4x^2 + 3x$

_____

  b. $9x^4 + 3x - 6$ and $7x^2 + 4x$

_____

17. Find each sum.

  a. Add $5x^4 - 7x^3 + 9$ and $-3x^4 + 8x^3 - 7$

$(5x^4 - 7x^3 + 9) + (-3x^4 + 8x^3 - 7)$

$= 5x^4 - 3x^4 - 7x^3 + 8x^3 + 9 - 7$

$= 2x^4 + x^3 + 2$

17. Find each sum.

  a. Add $15x^3 - 5x + 3$ and
$-11x^3 + 6x + 9$

_____

Name: Date:
Instructor: Section:

b. $(5x^4 - 7x^2 + 6x) + (-3x^3 + 4x^2 - 7)$

$(5x^4 - 7x^2 + 6x) + (-3x^3 + 4x^2 - 7)$
$= 5x^4 - 3x^3 - 7x^2 + 4x^2 + 6x - 7$
$= 5x^4 - 3x^3 - 3x^2 + 6x - 7$

b. $(8x^2 - 6x + 4) + (7x^3 - 8x - 5)$

_____

18. Perform each subtraction.

    a. $(8x - 3) - (4x - 7)$

    Use the definition of subtraction and the distributive property.
    $(8x - 3) - (4x - 7) = (8x - 3) + [-(4x - 7)]$
    $= (8x - 3) + [-1(4x - 7)]$
    $= (8x - 3) + (-4x + 7)$
    $= 4x + 4$

18. Perform each subtraction.

    a. $(7x - 6) - (5x - 8)$

    _____

    b. Subtract $8x^3 - 5x^2 + 8$ from $9x^3 + 6x^2 - 7$.

    $(9x^3 + 6x^2 - 7) - (8x^3 - 5x^2 + 8)$
    $= (9x^3 + 6x^2 - 7) + (-8x^3 + 5x^2 - 8)$
    $= x^3 + 11x^2 - 15$

    b. $(7x^3 - 3x - 5) - (18x^3 + 4x - 6)$

    _____

19. Subtract by columns:
    $(15x^3 - 8x^2 + 3x - 7) - (9x^3 - 5x^2 - 6x + 4)$.

    Arrange like terms in columns.
    $15x^3 - 8x^2 + 3x - 7$
    $\underline{9x^3 - 5x^2 - 6x + 4}$

    Change all the signs in the second row, and then add.
    $15x^3 - 8x^2 + 3x - 7$
    $\underline{-9x^3 + 5x^2 + 6x - 4}$
    $6x^3 - 3x^2 + 9x - 11$

19. Subtract by columns:
    $(16x^3 - 7x^2 - 5x + 3)$
    $-(7x^3 - 8x^2 + 6x - 4)$

    _____

CHAPTER 1 Equations and Inequalities

Name: Date:
Instructor: Section:

20. Perform the indicated operations to simplify the expression
$$(5-2x+9x^2)-(7-5x+8x^2)+(6+3x-5x^2)$$
Rewrite, changing the subtraction to adding the opposite.
$$(5-2x+9x^2)-(7-5x+8x^2)+(6+3x-5x^2)$$
$$=(5-2x+9x^2)+(-7+5x-8x^2)+(6+3x-5x^2)$$
$$=(-2+3x+x^2)+(6+3x-5x^2)$$
$$=4+6x-4x^2$$

21. Add or subtract as indicated.

a. $(5a+3ab+b)+(7a-2ab-b)$

$(5a+3ab+b)+(7a-2ab-b)$
$=5a+3ab+b+7a-2ab-b$
$=12a+ab$

b. $(3x^2y+5xy+y^2)-(4x^2y+xy-3y^2)$

$(3x^2y+5xy+y^2)-(4x^2y+xy-3y^2)$
$=3x^2y+5xy+y^2-4x^2y-xy+3y^2$
$=-x^2y+4xy+4y^2$

22. Find each product.

a. $5x^2(7x+3)$

Use the distributive property.
$$5x^2(7x+3)=5x^2(7x)+5x^2(3)$$
$$=35x^3+15x^2$$

b. $-9n^4(6n^4+5n^3+7n-1)$

Use the distributive property.
$-9n^4(6n^4+5n^3+7n-1)$
$=-9n^4(6n^4)-9n^4(5n^3)-9n^4(7n)-9n^4(-1)$
$=-54n^8-45n^7-63n^5+9n^4$

20. Perform the indicated operations to simplify the expression
$$(10-7x+6x^2)-(5-11x+3x^2)$$
$$+(2+4x-7x^2)$$

_____

21. Add or subtract as indicated.

a. $(10a+6ab+2b)$
$+(3a-5ab-2b)$

_____

b. $(7x^2y+3xy+4y^2)$
$-(6x^2y-xy+4y^2)$

_____

22. Find each product.

a. $8x^3(4x+8)$

_____

b. $-7m^5(5m^3-6m^2+4m-1)$

_____

Name: Date:
Instructor: Section:

23. Multiply $(x^2+6)(5x^3-4x^2+3x)$.

Multiply each term of the second polynomial by each term of the first.

$(x^2+6)(5x^3-4x^2+3x)$
$= x^2(5x^3)+x^2(-4x^2)+x^2(3x)$
$\quad +6(5x^3)+6(-4x^2)+6(3x)$
$= 5x^5-4x^4+3x^3+30x^3-24x^2+18x$
$= 5x^5-4x^4+33x^3-24x^2+18x$

24. Multiply $(2x^3+7x^2+5x-1)(4x+6)$ vertically.

Write the polynomials vertically.

$\quad 2x^3+7x^2+5x-1$
$\quad \qquad\qquad\quad 4x+6$

Begin by multiplying each term in the top row by 6.

$\quad 2x^3\ +7x^2\ +5x\ -1$
$\quad \qquad\qquad\qquad 4x\ +6$
$\overline{\quad 12x^3+42x^2+30x-6\ }$

Now multiply each term in the top row by $4x$. Then add like terms.

$\quad 2x^3\ +7x^2\ +5x\ -1$
$\quad \qquad\qquad\qquad 4x\ +6$
$\overline{\quad 12x^3\ +42x^2+30x-6\ }$
$8x^4+28x^3+20x^2\ -4x$
$\overline{8x^4+40x^3+62x^2+26x-6}$

The product is $8x^4+40x^3+62x^2+26x-6$.

23. Multiply
$(x^3+9)(4x^4-2x^2+x)$

_____

24. Multiply
$(4x^3-3x^2+6x+5)(7x-3)$
vertically.

_____

CHAPTER 1 Equations and Inequalities

Name: Date:
Instructor: Section:

25. Use the FOIL method to find the product $(x+7)(x-5)$.

   Step 1 F Multiply the first terms: $x(x) = x^2$.

   Step 2 O Find the outer product: $x(-5) = -5x$.

   Step 3 I Find the inner product: $7(x) = 7x$.
   Add the outer and inner products mentally:
   $$-5x + 7x = 2x$$

   Step 4 L Multiply the last terms: $7(-5) = -35$.

   The product $(x+7)(x-5)$ is $x^2 + 2x - 35$.

26. Multiply $(7x-3)(4y+5)$.

   First $7x(4y) = 28xy$

   Outer $7x(5) = 35x$

   Inner $-3(4y) = -12y$

   Last $-3(5) = -15$

   The product $(7x-3)(4y+5)$ is
   $28xy + 35x - 12y - 15$.

27. Find each product.

   a. $(3k+7m)(2k+9m)$

   $(3k+7m)(2k+9m)$
   $= 3k(2k) + 3k(9m) + 7m(2k) + 7m(9m)$
   $= 6k^2 + 27km + 14km + 63m^2$
   $= 6k^2 + 41km + 63m^2$

   b. $(8p-5q)(8p+q)$

   $(8p-5q)(8p+q)$
   $= 64p^2 + 8pq - 40pq - 5q^2$
   $= 64p^2 - 32pq - 5q^2$

25. Use the FOIL method to find the product $(x+9)(x-6)$.

   _____

26. Multiply $(8y-7)(2x+9)$.

   _____

27. Find each product.

   a. $(5k+8n)(3k+4n)$

   _____

   b. $(9p+4q)(5p-q)$

   _____

Copyright © 2015 Pearson Education, Inc.

Name: Date:
Instructor: Section:

c. $3x^2(x-5)(4x+7)$

$3x^2(x-5)(4x+7)$
$= 3x^2(4x^2-13x-35)$
$= 12x^4 - 39x^3 - 105x^2$

c. $5x^3(x-6)(7x+3)$

_____

Practice Problems

For extra help for exercises 1–3, see the videos on finding square roots in your MyMathLab course.

Find all square roots of each number.

1. 625

1. _____

2. $\dfrac{121}{196}$

2. _____

Find the square root

3. $\sqrt{\dfrac{900}{49}}$

3. _____

For extra help for exercises 4–6, see the videos on rational and irrational numbers in your MyMathLab course.

Tell whether each square root is rational, irrational, or not a real number.

4. $\sqrt{72}$

4. _____

5. $\sqrt{-36}$

5. _____

6. $\sqrt{6400}$

6. _____

For extra help for exercises 7–9, see the videos on simplifying radicals in your MyMathLab course.

Simplify. Assume that variables represent positive real numbers.

7. $\sqrt{8x^3y^6z^{11}}$

7. _____

CHAPTER 1 Equations and Inequalities

Name: 	Date:
Instructor: 	Section:

8. $\sqrt{\dfrac{5a^2b^3}{6}}$ 	8. _____

9. $\sqrt{\dfrac{7y^2}{12b}}$ 	9. _____

For extra help for exercise 10, see the videos on adding and subtracting radicals in your MyMathLab course.

Add or subtract. Assume that all variables represent positive real numbers.

10. $\sqrt{100x} - \sqrt{9x} + \sqrt{25x}$ 	10. _____

For extra help for exercises 11–12, see the videos on multiplying radicals in your MyMathLab course.

Multiply each product, then simplify.

11. $(\sqrt{5} + \sqrt{6})(\sqrt{2} - 4)$ 	11. _____

12. $(\sqrt{2} - \sqrt{12})^2$ 	12. _____

Name: Date:
Instructor: Section:

For extra help for exercises 13–15, see the videos on the product rule for exponents in your MyMathLab course.

Use the product rule to simplify each expression, if possible. Write each answer in exponential form.

13. $7^4 \cdot 7^3$ 13. _____

14. $(-2c^7)(-4c^8)$ 14. _____

15. $(3k^7)(-8k^2)(-2k^9)$ 15. _____

For extra help for exercises 16–18, see the videos on the zero exponent rule in your MyMathLab course.

Evaluate each expression.

16. -12^0 16. _____

17. $-15^0 - (-15)^0$ 17. _____

18. $\dfrac{0^8}{8^0}$ 18. _____

For extra help for exercises 19–21, see the videos on the negative exponent rule in your MyMathLab course.

Evaluate or simplify each expression, and write it using only positive exponents. Assume that all variables represent nonzero real numbers.

19. $-2k^{-4}$ 19. _____

20. $(m^2 n)^{-9}$ 20. _____

21. $\dfrac{2x^{-4}}{3y^{-7}}$ 21. _____

Copyright © 2015 Pearson Education, Inc.

CHAPTER 1 Equations and Inequalities

Name: Date:
Instructor: Section:

For extra help for exercises 22–30, see the videos on adding and subtracting polynomials in your MyMathLab course.

Add.

22. $\quad 9m^3 + 4m^2 - 2m + 3$
    $\quad \underline{-4m^3 - 6m^2 - 2m + 1}$

22. _____

23. $(x^2 + 6x - 8) + (3x^2 - 10)$

23. _____

24. $(3r^3 + 5r^2 - 6) + (2r^2 - 5r + 4)$

24. _____

Subtract.

25. $(-8w^3 + 11w^2 - 12) - (-10w^2 + 3)$

25. _____

26. $(8b^4 - 4b^3 + 7) - (2b^2 + b + 9)$

26. _____

Copyright © 2015 Pearson Education, Inc.

Name: Date:
Instructor: Section:

27. $(9x^3 + 7x^2 - 6x + 3) - (6x^3 - 6x + 1)$ 27. _____

Add or subtract as indicated.

28. $(-2a^6 + 8a^4b - b^2) - (a^6 + 7a^4b + 2b^2)$ 28. _____

29. $(4ab + 2bc - 9ac) + (3ca - 2cb - 9ba)$ 29. _____

30. $(2x^2y + 2xy - 4xy^2) + (6xy + 9xy^2) - (9x^2y + 5xy)$ 30. _____

For extra help for exercises 31–39, see the videos on multiplying polynomials in your MyMathLab course.

Find each product.

31. $7z(5z^3 + 2)$ 31. _____

Name: Date:
Instructor: Section:

32. $2m(3+7m^2+3m^3)$ 32. _____

33. $-3y^2(2y^3+3y^2-4y+11)$ 33. _____

34. $(x+3)(x^2-3x+9)$ 34. _____

35. $(2m^2+1)(3m^3+2m^2-4m)$ 35. _____

36. $(3x^2+x)(2x^2+3x-4)$ 36. _____

37. $(5a-b)(4a+3b)$ 37. _____

Name: Date:
Instructor: Section:

38. $(3+4a)(1+2a)$ **38.** _____

39. $(2m+3n)(-3m+4n)$ **39.** _____

Chapter 1 Equations and Inequalities

1.4R, 1.5R Factoring Out the Greatest Common Factor; Factoring Trinomials; Factoring Binomials

Key Terms

Use the vocabulary terms listed below to complete each statement in exercises 1–7.

    factor   factored form   greatest common factor (GCF)

    factoring   prime polynomial   perfect square trinomial

    difference

1. The process of writing a polynomial as a product is called _____.

2. An expression is in _____ when it is written as a product.

3. The _____ is the largest quantity that is a factor of each of a group of quantities.

4. An expression A is a _____ of an expression B if B can be divided by A with 0 remainder.

5. A _____ is a polynomial that cannot be factored using only integers.

6. A _____ is the result of a subtraction.

7. A _____ is a trinomial that can be factored as the square of a binomial.

Name: Date:
Instructor: Section:

Guided Examples

Review these examples:

1. Write in factored form by factoring out the greatest common factor.

   a. $7a^2 + 14a$

   GCF = $7a$
   $$7a^2 + 14a = 7a(a) + 7a(2)$$
   $$= 7a(a+2)$$

   Check Multiply the factored form.
   $$7a(a+2) = 7a(a) + 7a(2)$$
   $$= 7a^2 + 14a$$

   b. $12x^5 + 27x^4 - 15x^3$

   GCF = $3x^3$
   $$12x^5 + 27x^4 - 15x^3$$
   $$= 3x^3(4x^2) + 3x^3(9x) + 3x^3(-5)$$
   $$= 3x^3(4x^2 + 9x - 5)$$

   Check Multiply the factored form.
   $$3x^3(4x^2 + 9x - 5)$$
   $$= 3x^3(4x^2) + 3x^3(9x) + 3x^3(-5)$$
   $$= 12x^5 + 27x^4 - 15x^3$$

   c. $y^7 + y^4$

   GCF = y^4
   $$y^7 + y^4 = y^4(y^3) + y^4(1)$$
   $$= y^4(y^3 + 1)$$

   Check mentally by distributing y^4 over each term inside the parentheses.

Now Try:

1. Write in factored form by factoring out the greatest common factor.

   a. $8x^5 + 24x$

   _____

   b. $20y^4 - 12y^3 + 4y^2$

   _____

   c. $n^8 + n^7$

   _____

CHAPTER 1 Equations and Inequalities

Name: Date:
Instructor: Section:

d. $40y^5z^3 - 56y^3z^5$

GCF $= 8y^3z^3$

$40y^5z^3 - 56y^3z^5 = 8y^3z^3(5y^2) - 8y^3z^3(7z^2)$
$= 8y^3z^3(5y^2 - 7z^2)$

Check mentally by distributing $8y^3z^3$ over each term inside the parentheses.

2. Write $-15x^5 + 35x^4 - 5x^2$ in factored form.

$-5x^2$ is a common factor.
$-15x^5 + 35x^4 - 5x^2$
$= -5x^2(3x^3) - 5x^2(-7x^2) - 5x^2(1)$
$= -5x^2(3x^3 - 7x^2 + 1)$

Check

$-5x^2(3x^3 - 7x^2 + 1)$
$= -5x^2(3x^3) - 5x^2(-7x^2) - 5x^2(1)$
$= -15x^5 + 35x^4 - 5x^2$

3. Write in factored form by factoring out the greatest common factor.

a. $x(x+9) + 7(x+9)$

Factor out $x+9$.
$x(x+9) + 7(x+9) = (x+9)(x+7)$

b. $a^2(a+6) - 7(a+6)$

Factor out $a+6$.
$a^2(a+6) - 7(a+6) = (a+6)(a^2 - 7)$

d. $64x^5 - 40x^4 + 8x^3$

_____

2. Write $-9p^2q^3 - 27p^4q^2 + 9pq$ in factored form.

_____

3. Write in factored form by factoring out the greatest common factor.

a. $y(y+8) + 4(y+8)$

_____

b. $z^2(z+5) - 11(z+5)$

_____

Name: Date:
Instructor: Section:

4. Factor $m^2 + 8m + 15$.

   Look for integers whose product is 15 and whose sum is 8. Only positive signs are needed.

Factors of 15	Sums of Factors
15, 1	$15 + 1 = 16$
5, 3	$5 + 3 = 8$

   From the table, 5 and 3 are the required integers.
   $m^2 + 8m + 15$ factors as $(m+5)(m+3)$

   Check Use the FOIL method.
   $$(m+5)(m+3) = m^2 + 3m + 5m + 15$$
   $$= m^2 + 8m + 15$$

4. Factor $x^2 + 11x + 24$.

   _____

5. Factor $x^2 - 11x + 28$.

   Look for integers whose product is 28 and whose sum is −11. Since the numbers have a positive product and a negative sum, we consider only pairs of negative integers.

Factors of 28	Sums of Factors
−28, −1	$-28 + (-1) = -29$
−14, −2	$-14 + (-2) = -16$
−7, −4	$-7 + (-4) = -11$

   The required integers are −7 and −4.
   $x^2 - 11x + 28$ factors as $(x-7)(x-4)$

   Check Use the FOIL method.
   $$(x-7)(x-4) = x^2 - 4x - 7x + 28$$
   $$= x^2 - 11x + 28$$

5. Factor $y^2 - 12y + 35$.

   _____

CHAPTER 1 Equations and Inequalities

Name: Date:
Instructor: Section:

6. Factor $x^2 + 2x - 15$.

Look for integers whose product is –15 and whose sum is 2. To get a negative product, the pairs of integers must have different signs.

Factors of -15	Sums of Factors
15, –1	$15 + (-1) = 14$
–15, 1	$-15 + 1 = -14$
5, –3	$5 + (-3) = 2$

The required integers are 5 and –3.

$x^2 + 2x - 15$ factors as $(x+5)(x-3)$

Check Use the FOIL method.

$(x+5)(x-3) = x^2 - 3x + 5x - 15$
$= x^2 + 2x - 15$

7. Factor $x^2 - 10x - 39$.

Look for integers whose product is –39 and whose sum is –10. Because the constant term, –39, is negative, we need pairs of integers with different signs.

Factors of -39	Sums of Factors
39, –1	$39 + (-1) = 38$
–39, 1	$-39 + 1 = -38$
3, –13	$3 + (-13) = -10$

The required integers are –13 and 3.

$x^2 - 10x - 39$ factors as $(x-13)(x+3)$

Check Use the FOIL method.

$(x-13)(x+3) = x^2 + 3x - 13x - 39$
$= x^2 - 10x - 39$

6. Factor $p^2 + 6p - 27$.

_____

7. Factor $a^2 - 15a - 34$.

_____

Name: Date:
Instructor: Section:

8. Factor each trinomial.

   a. $x^2 - 7x + 18$

   Look for integers whose product is 18 and whose sum is –7. Since the numbers have a positive product and a negative sum, we consider only pairs of negative integers.

Factors of 18	Sums of Factors
$-18, -1$	$-18 + (-1) = -19$
$-9, -2$	$-9 + (-2) = -11$
$-6, -3$	$-6 + (-3) = -9$

   None of the pairs of integers has a sum of –7.
   $x^2 - 7x + 18$ cannot be factored.
   It is a prime polynomial.

   b. $k^2 - 5x + 13$

   There are no pairs of integers whose product is 13 and whose sum is –5, so $k^2 - 5x + 13$ is a prime polynomial.

9. Factor $x^2 - 6xy - 7y^2$.

   Here, the coefficient of x in the middle term is –6y, so we need to find two expressions whose product is $-7y^2$ and whose sum is –6y.

Factors of $-7y^2$	Sums of Factors
$7y, -y$	$7y + (-y) = 6y$
$-7y, y$	$-7y + y = -6y$

   $x^2 - 6xy - 7y^2$ factors as $(x - 7y)(x + y)$

   Check   Use the FOIL method.
$$(x - 7y)(x + y) = x^2 + xy - 7xy + 7y^2$$
$$= x^2 - 6xy - 7y^2$$

8. Factor each trinomial.

   a. $m^2 - 7m + 5$

   _____

   b. $x^2 - 3x + 10$

   _____

9. Factor $p^2 - 5pq - 14q^2$.

   _____

CHAPTER 1 Equations and Inequalities

Name: Date:
Instructor: Section:

15. Factor each trinomial.

   a. $x^2 - 26x + 169$

   The first and last terms are perfect squares. Check to see if the middle term is twice the product of the first and last terms of the binomial $x - 13$.
   $$2 \cdot x \cdot (-13) = -26x$$
   Thus, $x^2 - 26x + 169$ is a perfect square trinomial.
   $x^2 - 26x + 169$ factors as $(x-13)^2$.

   b. $49m^2 - 70m + 25$

   $49m^2 - 70m + 25$
   $= (7m)^2 + 2(7m)(-5) + (-5)^2$
   $= (7m - 5)^2$

   c. $36y^2 + 42y + 49$

   The first and last terms are perfect squares.
   $36y^2 = (6y)^2$ and $49 = 7^2$
   Twice the product of the first and last terms of the binomial is $2 \cdot (6y)(7) = 84y$, which is not the middle term of $36y^2 + 42y + 49$.

   It is a prime polynomial.

   d. $128z^3 + 192z^2 + 72z$

   $128z^3 + 192z^2 + 72z$
   $= 8z(16z^2 + 24z + 9)$
   $= 8z[(4z)^2 + 2(4z)(3) + 3^2]$
   $= 8z(4z + 3)^2$

15. Factor each trinomial.

   a. $x^2 - 24x + 144$

   _____

   b. $64m^2 + 48m + 9$

   _____

   c. $100y^2 - 70y + 49$

   _____

   d. $20x^3 + 100x^2 + 125x$

   _____

Name: Date:
Instructor: Section:

16. Factor each difference of cubes.

   a. $m^3 - 1000$

   Use the pattern for a difference of cubes.
   $$m^3 - 1000 = m^3 - 10^3$$
   $$= (m-10)(m^2 + 10m + 100)$$

   b. $64p^3 - 125$

   $$64p^3 - 125 = (4p^3) - 5^3$$
   $$= (4p-5)\left[(4p)^2 + (4p)(5) + 5^2\right]$$
   $$= (4p-5)(16p^2 + 20p + 25)$$

   c. $5m^3 - 135n^3$

   $$5m^3 - 135n^3 = 5(m^3 - 27n^3)$$
   $$= 5\left[m^3 - (3n)^3\right]$$
   $$= 5(m-3n)\left[m^2 + m(3n) + (3n)^2\right]$$
   $$= 5(m-3n)(m^2 + 3mn + 9n^2)$$

17. Factor each sum of cubes.

   a. $k^3 + 1000$

   $$k^3 + 1000 = k^3 + 10^3$$
   $$= (k+10)(k^2 - 10k + 100)$$

   b. $2m^3 + 250n^3$

   $$2m^3 + 250n^3 = 2(m^3 + 125n^3)$$
   $$= 2\left[m^3 + (5n)^3\right]$$
   $$= 2(m+5n)\left[m^2 - m(5n) + (5n)^2\right]$$
   $$= 2(m+5n)(m^2 - 5mn + 25n^2)$$

16. Factor each difference of cubes.

   a. $t^3 - 216$

   _____

   b. $27k^3 - y^3$

   _____

   c. $3x^3 - 192$

   _____

17. Factor each sum of cubes.

   a. $216x^3 + 1$

   _____

   b. $6x^3 + 48y^3$

   _____

Copyright © 2015 Pearson Education, Inc.

CHAPTER 1 Equations and Inequalities

Name: Date:
Instructor: Section:

Practice Problems

For extra help for exercises 1–3, see the videos on factoring out the greatest common factor in your MyMathLab course.

Factor out the greatest common factor or a negative common factor if the coefficient of the term of greatest degree is negative.

1. $20x^2 + 40x^2y - 70xy^2$ 1. _____

2. $2a(x-2y) + 9b(x-2y)$ 2. _____

3. $26x^8 - 13x^{12} + 52x^{10}$ 3. _____

For extra help for exercises 4–9, see the videos on properties of real numbers in your MyMathLab course.

Factor completely. If a polynomial cannot be factored, write prime.

4. $r^2 + r + 3$ 4. _____

5. $x^2 - 11x + 28$ 5. _____

6. $x^2 - 8x - 33$ 6. _____

Copyright © 2015 Pearson Education, Inc.

Name: Date:
Instructor: Section:

7. $2n^4 - 16n^3 + 30n^2$ 7. _____

8. $2a^3b - 10a^2b^2 + 12ab^3$ 8. _____

9. $10k^6 + 70k^5 + 100k^4$ 9. _____

For extra help for exercises 10–12, see the videos on factoring a difference of squares in your MyMathLab course.

Factor each binomial completely. If a binomial cannot be factored, write **prime**.

10. $x^2 - 49$ 10. _____

11. $81x^4 - 16$ 11. _____

12. $9x^2 + 16$ 12. _____

For extra help for exercises 13–15, see the videos on factoring perfect square trinomials in your MyMathLab course.

Factor each trinomial completely. It may be necessary to factor out the greatest common factor first.

13. $z^2 - \frac{4}{3}z + \frac{4}{9}$ 13. _____

14. $9j^2 + 12j + 4$ 14. _____

15. $-12a^2 + 60ab - 75b^2$ 15. _____

For extra help for exercises 16–18, see the videos on factoring a difference of cubes in your MyMathLab course.

Factor.

16. $8a^3 - 125b^3$ 16. _____

17. $216x^3 - 8y^3$ 17. _____

18. $(m+n)^3 - (m-n)^3$ 18. _____

For extra help for exercises 19–21, see the videos on factoring a sum of cubes in your MyMathLab course.

Factor.

19. $27r^3 + 8s^3$ 19. _____

20. $8a^3 + 64b^3$ 20. _____

21. $64x^3 + 343y^3$ 21. _____

Name: Date:
Instructor: Section:

Chapter 1 Equations and Inequalities

1.6R Rational Expressions; Lowest Terms of a Rational Expression; Operations with Rational Expressions; Factoring (including by substitution); Negative and Rational Exponents; Simplifying Radicals with Index Greater than 2

Key Terms

Use the vocabulary terms listed below to complete each statement in exercises 1–10.

rational expression	rational function	quadratic in form
standard form	product rule for exponents	
quotient rule for exponents		power rule for exponents
index	index (order)	radicand
cube root		

1. A _____ is a function that is defined by rational expression in the form $f(x) = \dfrac{P(x)}{Q(x)}$, where $Q(x) \neq 0$.

2. The quotient of two polynomials with denominator not 0 is called a _____.

3. A quadratic equation written in the form $ax^2 + bx + c = 0$, $a \neq 0$ is written in _____.

4. A nonquadratic equation that can be written as a quadratic equation is called _____.

5. $(x^2 y^3)^4 = x^8 y^{12}$ is an example of the _____.

6. $w^5 w^3 = w^8$ is an example of the _____.

7. $\dfrac{z^6}{z^4} = z^2$ is an example of the _____.

8. In a radical of the form $\sqrt[n]{a}$, the number n is the _____.

9. The number b is a _____ of a if $b^3 = a$.

10. In the expression $\sqrt[4]{x^2}$, the "4" is the _____ and x^2 is the _____.

CHAPTER 1 Equations and Inequalities

Name: _____ Date: _____
Instructor: _____ Section: _____

Guided Examples

Review these examples:	Now Try:
1. For each rational expression, find all numbers that are not in the domain. Then give the domain, using set-builder notation.	1. For each rational expression, find all numbers that are not in the domain. Then give the domain, using set-builder notation.

a. $\dfrac{5}{3x-9}$

Set the denominator equal to 0 and solve.
$$3x - 9 = 0$$
$$3x = 9$$
$$x = 3$$
The number 3 cannot be used for a replacement for x. The domain of f includes all real numbers except 3, written using set-builder notation as $\{x \mid x \neq 3\}$.

a. $\dfrac{9}{5x-10}$

_____

_____

b. $\dfrac{x}{x^2 - 3x + 2}$

Set the denominator equal to 0.
$$x^2 - 3x + 2 = 0$$
$$(x-1)(x-2) = 0$$
$$x - 1 = 0 \quad \text{or} \quad x - 2 = 0$$
$$x = 1 \quad \text{or} \quad x = 2$$
The domain of g includes all real numbers except 1 and 2, written $\{x \mid x \neq 1, \ 2\}$.

b. $\dfrac{x+2}{x^2 - 5x + 6}$

_____

_____

c. $\dfrac{2x - 6}{5}$

The denominator, 5, can never be 0, so the domain of h includes all real numbers, written in set-builder notation as $\{x \mid x \text{ is a real number}\}$.

c. $\dfrac{5x + 3}{7}$

_____

_____

d. $\dfrac{x-6}{x^2+1}$

Setting x^2+1 equal to 0 leads to $x^2=-1$. There is no real number whose square is -1. Therefore, any real number can be used as a replacement for x. The domain of f is $\{x \mid x \text{ is a real number}\}$.

2. Write each rational expression in lowest terms.

   a. $\dfrac{13x}{52}$

   $\dfrac{13x}{52} = \dfrac{x \cdot 13}{4 \cdot 13} = \dfrac{x}{4} \cdot 1 = \dfrac{x}{4}$

   b. $\dfrac{2x-6}{5}$

   This expression cannot be simplified further and is in lowest terms.

   c. $\dfrac{x^2-25}{x^2+10x+25}$

   $\dfrac{x^2-25}{x^2+10x+25} = \dfrac{(x+5)(x-5)}{(x+5)(x+5)} = \dfrac{x-5}{x+5}$

   d. $\dfrac{m^2-5}{3m+15}$

   $\dfrac{m^2-5}{3m+15} = \dfrac{(m+5)(m-5)}{3(m+5)} = \dfrac{m-5}{3}$

   e. $\dfrac{x^3-8}{x-2}$

   $\dfrac{x^3-8}{x-2} = \dfrac{(x-2)(x^2+2x+4)}{x-2} = x^2+2x+4$

d. $\dfrac{x-3}{x^2+9}$

2. Write each rational expression in lowest terms.

   a. $\dfrac{15m}{45}$

   b. $\dfrac{5x+3}{7}$

   c. $\dfrac{3y^2-13y-10}{2y^2-9y-5}$

   d. $\dfrac{x^2-9}{4x+12}$

   e. $\dfrac{x^3-1}{x-1}$

CHAPTER 1 Equations and Inequalities

Name: Date:
Instructor: Section:

f. $\dfrac{12xy - 28x - 15y + 35}{6xy - 14x + 3y - 7}$

$\dfrac{12xy - 28x - 15y + 35}{6xy - 14x + 3y - 7}$

$= \dfrac{(12xy - 15y) + (-28x + 35)}{(6xy + 3y) + (-14x - 7)}$

$= \dfrac{3y(4x - 5) - 7(4x - 5)}{3y(2x + 1) - 7(2x + 1)}$

$= \dfrac{(3y - 7)(4x - 5)}{(3y - 7)(2x + 1)}$

$= \dfrac{4x - 5}{2x + 1}$

f. $\dfrac{6xy + 4y - 3x - 2}{8xy - 6y - 4x + 3}$

_____

3. Write each rational expression in lowest terms.

   a. $\dfrac{p - 25}{25 - p}$

   $\dfrac{p - 25}{25 - p} = \dfrac{p - 25}{-1(p - 25)} = \dfrac{1}{-1} = -1$

   b. $\dfrac{x^2 - 49}{7 - x}$

   $\dfrac{x^2 - 49}{7 - x} = \dfrac{(x + 7)(x - 7)}{-1(x - 7)}$

   $= \dfrac{x + 7}{-1}$

   $= -1(x + 7)$, or $-x - 7$

3. Write each rational expression in lowest terms.

   a. $\dfrac{w - 6}{6 - w}$

   _____

   b. $\dfrac{x^2 - 100}{10 - x}$

   _____

4. Add or subtract as indicated.

   a. $\dfrac{7y}{9} + \dfrac{2x}{9}$

   Add the numerators. Keep the common denominator.

   $\dfrac{7y}{9} + \dfrac{2x}{9} = \dfrac{7y + 2x}{9}$

4. Add or subtract as indicated.

   a. $\dfrac{5y}{11} + \dfrac{7z}{11}$

   _____

Copyright © 2015 Pearson Education, Inc.

Name: Date:
Instructor: Section:

b. $\dfrac{13}{3k^2} - \dfrac{19}{3k^2}$

$\dfrac{13}{3k^2} - \dfrac{19}{3k^2} = \dfrac{13-19}{3k^2} = \dfrac{-6}{3k^2}$, or $-\dfrac{2}{k^2}$

c. $\dfrac{a}{a^2-b^2} + \dfrac{b}{a^2-b^2}$

$\dfrac{a}{a^2-b^2} + \dfrac{b}{a^2-b^2} = \dfrac{a+b}{a^2-b^2}$

$= \dfrac{a+b}{(a+b)(a-b)}$

$= \dfrac{1}{a-b}$

d. $\dfrac{5}{x^2+4x-5} + \dfrac{x}{x^2+4x-5}$

$= \dfrac{5+x}{x^2+4x-5}$

$= \dfrac{5+x}{(x+5)(x-1)}$

$= \dfrac{1}{x-1}$

5. Add or subtract as indicated.

a. $\dfrac{2}{5r} + \dfrac{3}{10r}$

$\dfrac{2}{5r} + \dfrac{3}{10r} = \dfrac{2 \cdot 2}{5r \cdot 2} + \dfrac{3}{10r}$

$= \dfrac{4}{10r} + \dfrac{3}{10r} = \dfrac{7}{10r}$

b. $\dfrac{4}{x} - \dfrac{3}{x+4}$

$\dfrac{4}{x} - \dfrac{3}{x+4} = \dfrac{4(x+4)}{x(x+4)} - \dfrac{3x}{x(x+4)}$

$= \dfrac{4x+16}{x(x+4)} - \dfrac{3x}{x(x+4)}$

$= \dfrac{4x+16-3x}{x(x+4)}$

$= \dfrac{x+16}{x(x+4)}$

b. $\dfrac{16}{5z^2} - \dfrac{26}{5z^2}$

_____

c. $\dfrac{c}{c^2-d^2} - \dfrac{d}{c^2-d^2}$

_____

d. $\dfrac{7}{x^2+6x-7} + \dfrac{x}{x^2+6x-7}$

_____

5. Add or subtract as indicated.

a. $\dfrac{7}{9z} + \dfrac{5}{18z}$

_____

b. $\dfrac{5}{s} - \dfrac{4}{s-6}$

_____

CHAPTER 1 Equations and Inequalities

Name:
Instructor:
Date:
Section:

6. Subtract.

   a. $\dfrac{19x}{5x+2} - \dfrac{4x-6}{5x+2}$

   $\dfrac{19x}{5x+2} - \dfrac{4x-6}{5x+2} = \dfrac{19x-(4x-6)}{5x+2}$

   $= \dfrac{19x-4x+6}{5x+2}$

   $= \dfrac{15x+6}{5x+2}$

   $= \dfrac{3(5x+2)}{5x+2}$

   $= 3$

   b. $\dfrac{1}{x-5} - \dfrac{1}{x+5}$

   $\dfrac{1}{x-5} - \dfrac{1}{x+5} = \dfrac{1(x+5)}{(x-5)(x+5)} - \dfrac{1(x-5)}{(x-5)(x+5)}$

   $= \dfrac{x+5-(x-5)}{(x-5)(x+5)}$

   $= \dfrac{x+5-x+5}{(x-5)(x+5)}$

   $= \dfrac{10}{(x-5)(x+5)}$

7. Add.

   $\dfrac{6p}{p-4} + \dfrac{2}{4-p}$

   Since the denominators are opposites, we multiply the second expression by −1.

   $= \dfrac{6p}{p-4} + \dfrac{2(-1)}{(4-p)(-1)}$

   $= \dfrac{6p}{p-4} + \dfrac{-2}{p-4}$

   $= \dfrac{6p-2}{p-4}$

6. Subtract.

   a. $\dfrac{23x}{6x+1} - \dfrac{5x-3}{6x+1}$

   _____

   b. $\dfrac{5}{x+3} - \dfrac{5}{x-3}$

   _____

7. Add.

   $\dfrac{x}{x-8} + \dfrac{12}{8-x}$

   _____

Name: Date:
Instructor: Section:

8. Add and subtract as indicated.
$$\frac{2}{x} - \frac{5}{x-2} + \frac{10}{x^2 - 2x}$$

$$= \frac{2}{x} - \frac{5}{x-2} + \frac{10}{x(x-2)}$$

$$= \frac{2(x-2)}{x(x-2)} - \frac{5x}{x(x-2)} + \frac{10}{x(x-2)}$$

$$= \frac{2x - 4 - 5x + 10}{x(x-2)}$$

$$= \frac{-3x + 6}{x(x-2)}$$

$$= \frac{-3(x-2)}{x(x-2)}$$

$$= -\frac{3}{x}$$

9. Subtract.
$$\frac{7x}{x^2 - 4x + 4} - \frac{2}{x^2 - 4}$$

$$= \frac{7x}{(x-2)^2} - \frac{2}{(x+2)(x-2)}$$

The LCD is $(x+2)(x-2)^2$.

$$= \frac{7x}{(x-2)^2} - \frac{2}{(x+2)(x-2)}$$

$$= \frac{7x(x+2)}{(x+2)(x-2)^2} - \frac{2(x-2)}{(x+2)(x-2)^2}$$

$$= \frac{7x(x+2) - 2(x-2)}{(x+2)(x-2)^2}$$

$$= \frac{7x^2 + 14x - 2x + 4}{(x+2)(x-2)^2}$$

$$= \frac{7x^2 + 12x + 4}{(x+2)(x-2)^2}$$

8. Add and subtract as indicated
$$\frac{12}{x^2 + 3x} - \frac{3}{x} + \frac{4}{x+3}$$

_____

9. Subtract.
$$\frac{8x}{x^2 - 10x + 25} - \frac{3}{x^2 - 25}$$

_____

CHAPTER 1 Equations and Inequalities

Name: Date:
Instructor: Section:

10. Add.

$$\frac{2q}{q^2-2q-8}+\frac{7}{3q^2-14q+8}$$

The LCD is $(q+2)(3q-2)(q-4)$.

$$=\frac{2q}{(q+2)(q-4)}+\frac{7}{(3q-2)(q-4)}$$

$$=\frac{2q(3q-2)}{(q+2)(3q-2)(q-4)}+\frac{7(q+2)}{(q+2)(3q-2)(q-4)}$$

$$=\frac{6q^2-4q+7q+14}{(q+2)(3q-2)(q-4)}$$

$$=\frac{6q^2+3q+14}{(q+2)(3q-2)(q-4)}$$

11. Multiply.

a. $\dfrac{7r-21}{r}\cdot\dfrac{3r^3}{8r-24}$

$$\frac{7r-21}{r}\cdot\frac{3r^3}{8r-24}=\frac{7(r-3)}{r}\cdot\frac{3r\cdot r}{8(r-3)}$$

$$=\frac{r(r-3)}{r(r-3)}\cdot\frac{7\cdot 3r}{8}$$

$$=\frac{21r}{8}$$

b. $\dfrac{w^2-w-2}{w^2-6w+8}\cdot\dfrac{w^2-4w}{w^2+4w+3}$

$$\frac{w^2-w-2}{w^2-6w+8}\cdot\frac{w^2-4w}{w^2+4w+3}$$

$$=\frac{(w-2)(w+1)}{(w-2)(w-4)}\cdot\frac{w(w-4)}{(w+3)(w+1)}$$

$$=\frac{w}{w+3}$$

10. Add.

$$\frac{5}{p^2-6p+5}+\frac{4}{p^2-1}$$

_____

11. Multiply.

a. $\dfrac{9x-36}{x}\cdot\dfrac{4x^2}{7x-28}$

_____

b. $\dfrac{m^2+3m}{m^2-2m-3}\cdot\dfrac{m^2-2m-3}{m^2+m-6}$

_____

Copyright © 2015 Pearson Education, Inc.

Name: Date:
Instructor: Section:

c. $(y-1) \cdot \dfrac{6}{7y-7}$

$(y-1) \cdot \dfrac{6}{7y-7} = \dfrac{y-1}{1} \cdot \dfrac{6}{7(y-1)}$

$= \dfrac{6}{7}$

d. $\dfrac{x^2-2x}{x+5} \cdot \dfrac{x^2+8x+15}{x^3-3x^2}$

$\dfrac{x^2-2x}{x+5} \cdot \dfrac{x^2+8x+15}{x^3-3x^2}$

$= \dfrac{x(x-2)}{x+5} \cdot \dfrac{(x+5)(x+3)}{x^2(x-3)}$

$= \dfrac{(x+3)(x-2)}{x(x-3)}$

e. $\dfrac{x+2}{x^2+4x+4} \cdot \dfrac{x^2+x-2}{x^2-3x+2}$

$\dfrac{x+2}{x^2+4x+4} \cdot \dfrac{x^2+x-2}{x^2-3x+2}$

$= \dfrac{x+2}{(x+2)^2} \cdot \dfrac{(x-1)(x+2)}{(x-1)(x-2)}$

$= \dfrac{1}{x-2}$

12. Divide.

a. $\dfrac{8x^2}{21} \div \dfrac{2x}{7}$

$\dfrac{8x^2}{21} \div \dfrac{2x}{7} = \dfrac{8x^2}{21} \cdot \dfrac{7}{2x} = \dfrac{4x \cdot 2x}{3 \cdot 7} \cdot \dfrac{7}{2x} = \dfrac{4x}{3}$

c. $(p-6) \cdot \dfrac{4}{5p-30}$

_____

d. $\dfrac{x^2-x}{x-3} \cdot \dfrac{x^2+x-12}{x^3+2x^2}$

_____

e. $\dfrac{x+4}{x^2+8x+16} \cdot \dfrac{x^2+3x-4}{x^2-4x+3}$

_____

12. Divide.

a. $\dfrac{10p}{21} \div \dfrac{15p^2}{7}$

_____

CHAPTER 1 Equations and Inequalities

Name: Date:
Instructor: Section:

b. $\dfrac{3x+7}{2x^3} \div \dfrac{6x+14}{10x^6}$

$\dfrac{3x+7}{2x^3} \div \dfrac{6x+14}{10x^6} = \dfrac{3x+7}{2x^3} \cdot \dfrac{10x^6}{6x+14}$

$\phantom{\dfrac{3x+7}{2x^3} \div \dfrac{6x+14}{10x^6}} = \dfrac{3x+7}{2x^3} \cdot \dfrac{5x^3 \cdot 2x^3}{2(3x+7)}$

$\phantom{\dfrac{3x+7}{2x^3} \div \dfrac{6x+14}{10x^6}} = \dfrac{5x^3}{2}$

b. $\dfrac{2x-5}{3x^2} \div \dfrac{6x-15}{15x^3}$

_____

c. $\dfrac{2k^2+5k-12}{2k^2+k-3} \div \dfrac{k^2+8k+16}{2k^2+k-3}$

$\dfrac{2k^2+5k-12}{2k^2+k-3} \div \dfrac{k^2+8k+16}{2k^2+k-3}$

$= \dfrac{2k^2+5k-12}{2k^2+k-3} \cdot \dfrac{2k^2+k-3}{k^2+8k+16}$

$= \dfrac{(2k-3)(k+4)}{(2k+3)(k-1)} \cdot \dfrac{(2k+3)(k-1)}{(k+4)(k+4)}$

$= \dfrac{2k-3}{k+4}$

c. $\dfrac{2a^2-5a-12}{a^2-10a+24} \div \dfrac{a^2-9}{a^2-9a+18}$

_____

13. Define a variable u and write each equation in the form $au^2+bu+c=0$.

a. $c^4-20c^2+64=0$

Because $c^4=(c^2)^2$, we can define $u=c^2$, and rewrite the original equation as a quadratic equation in u.
$u^2-20u+64=0$

b. $(m+5)^2+6(m+5)+8=0$

Because this equation involves both $(m+5)^2$ and $(m+5)$, we choose $u=m+5$. We rewrite the original equation as a quadratic equation in u.
$u^2+6u+8=0$

13. Define a variable u and write each equation in the form $au^2+bu+c=0$.

a. $x^4-5x^2+4=0$

_____

b. $(x-5)^2+2(x-5)-35=0$

_____

Name: Date:
Instructor: Section:

c. $x^{4/3} - 20x^{2/3} + 36 = 0$

Because $(x^{2/3})^2 = x^{4/3}$, we define $u = x^{2/3}$. We rewrite the original equation as a quadratic equation in u.

$u^2 - 20u + 36 = 0$

14. Solve each equation.

a. $c^4 - 20c^2 + 64 = 0$

Write this equation in quadratic form by substituting u for c^2.

$u^2 - 20u + 64 = 0$
$(u - 4)(u - 16) = 0$
$u - 4 = 0$ or $u - 16 = 0$
$u = 4$ or $u = 16$
$c^2 = 4$ or $c^2 = 16$
$c = \pm 2$ or $c = \pm 4$

The solution set is $\{-4, -2, 2, 4\}$.

b. $m^4 - 5m^2 + 4 = 0$

$m^4 - 5m^2 + 4 = 0$
$u^2 - 5u + 4 = 0$ Let $u = x^2$.
$(u - 4)(u - 1) = 0$
$u - 4 = 0$ or $u - 1 = 0$
$u = 4$ or $u = 1$
$m^2 = 4$ or $m^2 = 1$
$m = \pm 2$ or $m = \pm 1$

The solution set is $\{-2, -1, 1, 2\}$.

c. $x^{2/3} - 2x^{1/3} - 3 = 0$

_____

14. Solve each equation.

a. $x^4 - 5x^2 + 4 = 0$

_____

b. $16m^4 = 25m^2 - 9$

_____

Name: Date:
Instructor: Section:

c. $x^4 - 4x^2 + 1 = 0$

Let $u = x^2$. Then the equation becomes $u^2 - 4u + 1 = 0$. This equation cannot be solved by factoring, so we use the quadratic formula.

$$u = \frac{-(-4) \pm \sqrt{(-4)^2 - 4(1)(1)}}{2(1)}$$

$$u = \frac{4 \pm \sqrt{12}}{2} = \frac{4 \pm 2\sqrt{3}}{2} = 2 \pm \sqrt{3}$$

Now substitute and solve for x.

$u = 2 - \sqrt{3}$ or $u = 2 + \sqrt{3}$
$x^2 = 2 - \sqrt{3}$ or $x^2 = 2 + \sqrt{3}$
$x = \pm\sqrt{2 - \sqrt{3}}$ or $x = \pm\sqrt{2 + \sqrt{3}}$

The solution set is $\{-\sqrt{2-\sqrt{3}}, -\sqrt{2+\sqrt{3}}, \sqrt{2-\sqrt{3}}, \sqrt{2+\sqrt{3}}\}$.

c. $x^4 - 4x^2 - 8 = 0$

_____

15. Simplify by writing with positive exponents. Assume that all variables represent nonzero real numbers.

a. 5^{-2}

$5^{-2} = \frac{1}{5^2}$, or $\frac{1}{25}$

b. 4^{-3}

$4^{-3} = \frac{1}{4^3}$, or $\frac{1}{64}$

c. $\left(\frac{1}{3}\right)^{-3}$

$\left(\frac{1}{3}\right)^{-3} = 3^3$, or 27

15. Simplify by writing with positive exponents. Assume that all variables represent nonzero real numbers.

a. 8^{-2}

_____

b. 3^{-3}

_____

c. $\left(\frac{1}{5}\right)^{-2}$

_____

Name: Date:
Instructor: Section:

d. $\left(\dfrac{3}{5}\right)^{-4}$

$\left(\dfrac{3}{5}\right)^{-4} = \left(\dfrac{5}{3}\right)^{4}$ The reciprocal of $\dfrac{3}{5}$ is $\dfrac{5}{3}$.

$= \dfrac{5^4}{3^4}$

$= \dfrac{625}{81}$

d. $\left(\dfrac{6}{7}\right)^{-2}$

_____

e. $\left(\dfrac{2}{3}\right)^{-5}$

$\left(\dfrac{2}{3}\right)^{-5} = \left(\dfrac{3}{2}\right)^{5}$

$= \dfrac{3^5}{2^5}$

$= \dfrac{243}{32}$

e. $\left(\dfrac{3}{2}\right)^{-3}$

_____

f. $5^{-1} - 3^{-1}$

$5^{-1} - 3^{-1} = \dfrac{1}{5} - \dfrac{1}{3}$

$= \dfrac{3}{15} - \dfrac{5}{15}$

$= -\dfrac{2}{15}$

f. $4^{-1} - 8^{-1}$

_____

g. $4q^{-3}$

$4q^{-3} = \dfrac{4}{1} \cdot \dfrac{1}{q^3}$

$= \dfrac{4}{q^3}$

g. $11p^{-4}$

_____

h. $\dfrac{1}{x^{-6}}$

$\dfrac{1}{x^{-6}} = \dfrac{1^{-6}}{x^{-6}}$

$= \left(\dfrac{1}{x}\right)^{-6}$

$= x^6$

h. $\dfrac{1}{x^{-8}}$

_____

CHAPTER 1 Equations and Inequalities

Name: Date:
Instructor: Section:

i. $x^5 y^{-8}$

$$x^5 y^{-8} = \frac{x^5}{1} \cdot \frac{1}{y^8}$$

$$= \frac{x^5}{y^8}$$

i. $a^{-5} b^4$

_____

16. Simplify. Assume that all variables represent nonzero real numbers.

a. $\dfrac{3^{-4}}{7^{-2}} = \dfrac{7^2}{3^4}$, or $\dfrac{49}{81}$

b. $\dfrac{a^{-6}}{b^{-1}} = \dfrac{b^1}{a^6}$, or $\dfrac{b}{a^6}$

c. $\dfrac{x^{-3} y}{4 z^{-4}} = \dfrac{y z^4}{4 x^3}$

d. $\left(\dfrac{p}{3q} \right)^{-3}$

$\left(\dfrac{p}{3q} \right)^{-3} = \left(\dfrac{3q}{p} \right)^3$

$= \dfrac{3^3 q^3}{p^3}$

$= \dfrac{27 q^3}{p^3}$

16. Simplify. Assume that all variables represent nonzero real numbers.

a. $\dfrac{6^{-2}}{5^{-3}}$

_____

b. $\dfrac{x^{-7}}{y^{-1}}$

_____

c. $\dfrac{p^{-3} q}{4 r^{-5}}$

_____

d. $\left(\dfrac{a}{5b} \right)^{-4}$

_____

7. Evaluate each exponential.

a. $8^{1/3}$

$8^{1/3} = \sqrt[3]{8} = 2$

17. Evaluate each exponential.

a. $216^{1/3}$

_____

Name: Date:
Instructor: Section:

b. $25^{1/2}$

$25^{1/2} = \sqrt{25} = 5$

c. $-81^{1/4}$

$-81^{1/4} = -\sqrt[4]{81} = -3$

d. $(-81)^{1/4}$

$(-81)^{1/4} = \sqrt[4]{-81}$ is not a real number.

e. $(-8)^{1/3}$

$(-8)^{1/3} = \sqrt[3]{-8} = -2$

f. $\left(\dfrac{1}{125}\right)^{1/3}$

$\left(\dfrac{1}{125}\right)^{1/3} = \sqrt[3]{\dfrac{1}{125}} = \dfrac{1}{5}$

18. Evaluate each exponential.

a. $100^{3/2}$

$100^{3/2} = (100^{1/2})^3 = 10^3 = 1000$

b. $64^{2/3}$

$64^{2/3} = (64^{1/3})^2 = 4^2 = 16$

c. $-729^{5/6}$

$-729^{5/6} = -(729^{1/6})^5 = -(3)^5 = -243$

d. $(-125)^{2/3}$

$(-125)^{2/3} = [(-125)^{1/3}]^2 = (-5)^2 = 25$

b. $121^{1/2}$

_____

c. $-1024^{1/10}$

_____

d. $(-1024)^{1/10}$

_____

e. $(-243)^{1/5}$

_____

f. $\left(\dfrac{1}{16}\right)^{1/4}$

_____

18. Evaluate each exponential.

a. $27^{2/3}$

_____

b. $16^{3/4}$

_____

c. $-36^{3/2}$

_____

d. $(-27)^{2/3}$

_____

e. $(-729)^{5/6}$

$(-729)^{5/6} = [(-729)^{1/6}]^5$ is not a real number since $(-729)^{1/6}$ is not a real number.

19. Evaluate each exponential.

  a. $625^{-3/4}$

  $625^{-3/4} = \dfrac{1}{625^{3/4}} = \dfrac{1}{(625^{1/4})^3} = \dfrac{1}{\left(\sqrt[4]{625}\right)^3}$

  $= \dfrac{1}{5^3} = \dfrac{1}{125}$

  b. $36^{-3/2}$

  $36^{-3/2} = \dfrac{1}{36^{3/2}} = \dfrac{1}{\left(\sqrt{36}\right)^3} = \dfrac{1}{6^3} = \dfrac{1}{216}$

  c. $\left(\dfrac{81}{16}\right)^{-3/4}$

  $\left(\dfrac{81}{16}\right)^{-3/4} = \left(\dfrac{16}{81}\right)^{3/4} = \left(\sqrt[4]{\dfrac{16}{81}}\right)^3 = \left(\dfrac{2}{3}\right)^3 = \dfrac{8}{27}$

20. Write each radical as an exponential and simplify. Assume that all variables represent positive real numbers. Use the definition that takes the root first.

  a. $17^{1/2}$

  $17^{1/2} = \sqrt{17}$

  b. $10^{5/6}$

  $10^{5/6} = \left(\sqrt[6]{10}\right)^5$

  c. $8x^{3/4}$

  $8x^{3/4} = 8\left(\sqrt[4]{x}\right)^3$

e. $(-36)^{3/2}$

_____

19. Evaluate each exponential.

  a. $32^{-2/5}$

  _____

  b. $125^{-4/3}$

  _____

  c. $\left(\dfrac{27}{64}\right)^{-2/3}$

  _____

20. Write each radical as an exponential and simplify. Assume that all variables represent positive real numbers. Use the definition that takes the root first.

  a. $23^{1/3}$

  _____

  b. $21^{3/4}$

  _____

  c. $5x^{5/4}$

  _____

Name: Date:
Instructor: Section:

d. $2x^{2/5} - (4x)^{5/6}$

$2x^{2/5} - (4x)^{5/6} = 2\left(\sqrt[5]{x}\right)^2 - \left(\sqrt[6]{4x}\right)^5$

e. $x^{-4/5}$

$x^{-4/5} = \dfrac{1}{x^{4/5}} = \dfrac{1}{\left(\sqrt[4]{x}\right)^5}$

f. $\left(x^3 + y^2\right)^{1/5}$

$\left(x^3 + y^2\right)^{1/5} = \sqrt[5]{x^3 + y^2}$

21. Write each radical as an exponential and simplify. Assume that all variables represent positive real numbers.

a. $\sqrt{14}$

$\sqrt{14} = 14^{1/2}$

b. $\sqrt[3]{3^6}$

$\sqrt[3]{3^6} = 3^{6/3} = 3^2 = 9$

c. $\sqrt[7]{w^7}$

$\sqrt[7]{w^7} = w^{7/7} = w^1 = w$

22. Write with only positive exponents. Assume that all variables represent positive real numbers.

a. $13^{4/5} \cdot 13^{1/2}$

$13^{4/5} \cdot 13^{1/2} = 13^{4/5 + 1/2} = 13^{13/10}$

b. $\dfrac{8^{3/4}}{8^{1/4}}$

$\dfrac{8^{3/4}}{8^{1/4}} = 8^{3/4 - 1/4} = 8^{1/2}$

d. $(2x)^{4/3} - 3x^{2/5}$

_____

e. $x^{-3/2}$

_____

f. $\left(x^2 - y^2\right)^{1/4}$

_____

21. Write each radical as an exponential and simplify. Assume that all variables represent positive real numbers.

a. $\sqrt{22}$

_____

b. $\sqrt{10^4}$

_____

c. $\sqrt[8]{m^8}$

_____

22. Write with only positive exponents. Assume that all variables represent positive real numbers.

a. $5^{3/4} \cdot 5^{7/4}$

_____

b. $\dfrac{a^{4/5}}{a^{2/3}}$

_____

Copyright © 2015 Pearson Education, Inc.

Name: Date:
Instructor: Section:

c. $\dfrac{\left(n^{7/4}w^{1/2}\right)^2}{w^{3/4}}$

$\dfrac{\left(n^{7/4}w^{1/2}\right)^2}{w^{3/4}} = \dfrac{\left(n^{7/4}\right)^2\left(w^{1/2}\right)^2}{w^{3/4}}$

$= \dfrac{n^{7/2}w^1}{w^{3/4}}$

$= n^{7/2}w^{1-3/4}$

$= n^{7/2}w^{1/4}$

d. $\left(\dfrac{c^6 x^3}{c^{-2} x^{1/2}}\right)^{-3/4}$

$\left(\dfrac{c^6 x^{1/2}}{c^{-2} x^3}\right)^{-3/4} = \left(c^{6-(-2)} x^{1/2-3}\right)^{-3/4}$

$= \left(c^8 x^{-5/2}\right)^{-3/4}$

$= \left(c^8\right)^{-3/4}\left(x^{-5/2}\right)^{-3/4}$

$= c^{-6} x^{15/8}$

$= \dfrac{x^{15/8}}{c^6}$

e. $a^{3/4}(a^{2/3} - a^{1/2})$

$a^{3/4}(a^{2/3} - a^{1/2}) = a^{3/4}a^{2/3} - a^{3/4}a^{1/2}$

$= a^{3/4+2/3} - a^{3/4+1/2}$

$= a^{17/12} - a^{5/4}$

c. $\dfrac{\left(x^{1/3}y^{2/3}\right)^6}{y^{1/2}}$

_____

d. $\left(\dfrac{x^{-1}y^{2/3}}{x^{1/3}y^{1/2}}\right)^{-3/2}$

_____

e. $r^{1/2}(r^{2/3} - r^{8/3})$

_____

Name: Date:
Instructor: Section:

23. Write all radicals as exponentials, and then apply the rules for rational exponents. Leave answers in exponential form. Assume that all variables represent positive real numbers.

   a. $\sqrt[4]{x^3} \cdot \sqrt[5]{x}$

   $\sqrt[4]{x^3} \cdot \sqrt[5]{x} = x^{3/4} \cdot x^{1/5}$
   $= x^{3/4 + 1/5}$
   $= x^{15/20 + 4/20}$
   $= x^{19/20}$

   b. $\dfrac{\sqrt[3]{y^5}}{\sqrt{y^3}}$

   $\dfrac{\sqrt[3]{y^5}}{\sqrt{y^3}} = \dfrac{y^{5/3}}{y^{3/2}} = y^{5/3 - 3/2} = y^{1/6}$

   c. $\sqrt{\sqrt[4]{y^3}}$

   $\sqrt{\sqrt[4]{y^3}} = \sqrt{y^{3/4}} = \left(y^{3/4}\right)^{1/2} = y^{3/8}$

24. Find each cube root.

   a. $\sqrt[3]{64}$

   Because $4^3 = 64$, $\sqrt[3]{64} = 4$.

   b. $\sqrt[3]{-64}$

   $\sqrt[3]{-64} = -4$, because $(-4)^3 = -64$.

   c. $\sqrt[3]{729}$

   $\sqrt[3]{729} = 9$, because $9^3 = 729$.

23. Write all radicals as exponentials, and then apply the rules for rational exponents. Leave answers in exponential form. Assume that all variables represent positive real numbers.

   a. $\sqrt[6]{x^3} \cdot \sqrt[3]{x^2}$

   b. $\dfrac{\sqrt[4]{y^5}}{\sqrt[3]{y^2}}$

   c. $\sqrt[3]{\sqrt[4]{x^3}}$

24. Find each cube root.

   a. $\sqrt[3]{125}$

   b. $\sqrt[3]{-125}$

   c. $\sqrt[3]{343}$

Copyright © 2015 Pearson Education, Inc.

CHAPTER 1 Equations and Inequalities

Name: Date:
Instructor: Section:

25. Find each root.

   a. $\sqrt[4]{81}$

$\sqrt[4]{81} = 3$, because 3 is positive and $3^4 = 81$.

   b. $-\sqrt[4]{81}$

From part (a), $\sqrt[4]{81} = 3$, so the negative root is $-\sqrt[4]{81} = -3$.

   c. $\sqrt[4]{-81}$

For a real number fourth root, the radicand must be nonnegative. There is no real number that equals $\sqrt[4]{-81}$.

   d. $-\sqrt[5]{1024}$

$-\sqrt[5]{1024} = -4$

   e. $\sqrt[5]{-1024}$

$\sqrt[5]{-1024} = -4$, because $(-4)^5 = -1024$.

26. Simplify each root.

   a. $\sqrt[4]{(-5)^4}$

n is even. Use absolute value.

$\sqrt[4]{(-5)^4} = |-5| = 5$

   b. $\sqrt[5]{(-3)^5}$

n is odd.

$\sqrt[5]{(-3)^5} = -3$

   c. $-\sqrt[6]{(-8)^6}$

n is even. Use absolute value.

$-\sqrt[6]{(-8)^6} = -|-8| = -8$

25. Find each root.

   a. $\sqrt[4]{625}$

      _____

   b. $-\sqrt[4]{625}$

      _____

   c. $\sqrt[4]{-625}$

      _____

   d. $-\sqrt[5]{3125}$

      _____

   e. $\sqrt[5]{-3125}$

      _____

26. Simplify each root.

   a. $\sqrt[8]{(-4)^8}$

      _____

   b. $\sqrt[7]{(-5)^7}$

      _____

   c. $-\sqrt[4]{(-2)^4}$

      _____

Copyright © 2015 Pearson Education, Inc.

Name: Date:
Instructor: Section:

d. $-\sqrt{r^{12}}$

$-\sqrt{r^{12}} = -|r^6| = -r^6$

No absolute value bars are needed here since r^6 is nonnegative for any real number value of r.

e. $\sqrt[5]{s^{20}}$

$\sqrt[5]{s^{20}} = s^4$, because $s^{20} = (s^4)^5$.

f. $\sqrt[4]{x^{20}}$

$\sqrt[4]{x^{20}} = |x^5|$

27. Simplify.

a. $\sqrt[3]{\dfrac{16}{9}}$

$\sqrt[3]{\dfrac{16}{9}} = \dfrac{\sqrt[3]{8 \cdot 2}}{\sqrt[3]{9}} = \dfrac{2\sqrt[3]{2}}{\sqrt[3]{9}}$

$= \dfrac{2\sqrt[3]{2} \cdot \sqrt[3]{3}}{\sqrt[3]{9} \cdot \sqrt[3]{3}}$

$= \dfrac{2\sqrt[3]{6}}{\sqrt[3]{27}}$

$= \dfrac{2\sqrt[3]{6}}{3}$

b. $\sqrt[4]{\dfrac{28s}{r}}$, $r > 0$, $s \geq 0$

$\sqrt[4]{\dfrac{28s}{r}} = \dfrac{\sqrt[4]{28s}}{\sqrt[4]{r}}$

$= \dfrac{\sqrt[4]{28s} \cdot \sqrt[4]{r^3}}{\sqrt[4]{r} \cdot \sqrt[4]{r^3}}$

$= \dfrac{\sqrt[4]{28sr^3}}{r}$

d. $-\sqrt{x^8}$

_____

e. $\sqrt[3]{w^{30}}$

_____

f. $\sqrt[6]{x^{30}}$

_____

27. Simplify.

a. $\sqrt[3]{\dfrac{8}{100}}$

_____

b. $\sqrt[4]{\dfrac{4t}{x}}$, $t \geq 0$, $x > 0$

_____

Copyright © 2015 Pearson Education, Inc.

IRW-94 CHAPTER 1 Equations and Inequalities

Name: Date:
Instructor: Section:

28. Multiply. Assume that all variables represent positive real numbers.

   a. $\sqrt[4]{2} \cdot \sqrt[4]{2x}$

   $\sqrt[4]{2} \cdot \sqrt[4]{2x} = \sqrt[4]{2 \cdot 2x} = \sqrt[4]{4x}$

   b. $\sqrt[3]{8x} \cdot \sqrt[3]{2y^2}$

   $\sqrt[3]{8x} \cdot \sqrt[3]{2y^2} = \sqrt[3]{8x \cdot 2y^2} = \sqrt[3]{16xy^2}$

   c. $\sqrt[5]{6r^2} \cdot \sqrt[5]{4r^2}$

   $\sqrt[5]{6r^2} \cdot \sqrt[5]{4r^2} = \sqrt[5]{6r^2 \cdot 4r^2} = \sqrt[5]{24r^4}$

   d. $\sqrt[5]{2} \cdot \sqrt[4]{6}$

   $\sqrt[5]{2} \cdot \sqrt[4]{6}$ cannot be simplified using the product rule for radicals, because the indexes (5 and 4) are different.

   e. $(2 + \sqrt[3]{5})(2 - \sqrt[3]{5})$

   $(2 + \sqrt[3]{5})(2 - \sqrt[3]{5})$
   $= 2 \cdot 2 - 2\sqrt[3]{5} + 2\sqrt[3]{5} - \sqrt[3]{5} \cdot \sqrt[3]{5}$
   $= 4 - \sqrt[3]{5^2}$
   $= 4 - \sqrt[3]{25}$

28. Multiply. Assume that all variables represent positive real numbers.

   a. $\sqrt[3]{3} \cdot \sqrt[3]{7}$

   _____

   b. $\sqrt[3]{7x} \cdot \sqrt[3]{5y}$

   _____

   c. $\sqrt[5]{4w} \cdot \sqrt[5]{2w^3}$

   _____

   d. $\sqrt{3} \cdot \sqrt[3]{64}$

   _____

   e. $(4 - \sqrt[3]{2})(4 + \sqrt[3]{2})$

   _____

Practice Problems

For extra help for exercises 1–3, see the videos on finding the domain of a rational expression in your MyMathLab course.

Find all numbers that are not in the domain of each expression. Then give the domain using set notation.

1. $\dfrac{8s + 7}{3s - 2}$

1. _____

   _____

Name:
Instructor:

Date:
Section:

2. $\dfrac{x-6}{x^2+1}$

2. _____

_____

3. $\dfrac{q+7}{q^2-3q+2}$

3. _____

_____

For extra help for exercises 4–6, see the videos on writing rational expressions in lowest terms in your MyMathLab course.

Write each rational expression in lowest terms.

4. $\dfrac{12k^3+12k^2}{3k^2+3k}$

4. _____

5. $\dfrac{2y^2-3y-5}{2y^2-11y+15}$

5. _____

6. $\dfrac{a^2-3a}{3a-a^2}$

6. _____

For extra help for exercises 7–12, see the videos on adding and subtracting rational expressions in lowest terms in your MyMathLab course.

Add or subtract as indicated. Write each answer in lowest terms.

7. $\dfrac{n}{m+3}-\dfrac{-3n+7}{m+3}$

7. _____

Copyright © 2015 Pearson Education, Inc.

CHAPTER 1 Equations and Inequalities

Name:
Instructor:

Date:
Section:

8. $\dfrac{2x+3}{x^2+3x-10} + \dfrac{2-x}{x^2+3x-10}$

8. _____

9. $\dfrac{k}{k^2-6k+8} - \dfrac{2}{k^2-6k+8}$

9. _____

10. $\dfrac{3}{n^2-16} - \dfrac{6n}{n^2+8n+16}$

10. _____

11. $\dfrac{4z}{z^2+6z+8} + \dfrac{2z-1}{z^2+5z+6}$

11. _____

12. $\dfrac{4y}{y^2+4y+3} - \dfrac{3y+1}{y^2-y-2}$

12. _____

Name: Date:
Instructor: Section:

For extra help for exercises 13–15, see the videos on multiplying rational expressions in lowest terms in your MyMathLab course.

Multiply. Write each answer in lowest terms.

13. $\dfrac{x^2+x-12}{x^2+7x+10} \cdot \dfrac{x^2+3x-10}{x^2+2x-8}$

13. _____

14. $\dfrac{x^2+10x+21}{x^2+14x+49} \cdot \dfrac{x^2+12x+35}{x^2-6x-27}$

14. _____

15. $\dfrac{3m^2-m-10}{2m^2-7m-4} \cdot \dfrac{4m^2-1}{6m^2+7m-5}$

15. _____

For extra help for exercises 16–18, see the videos on multiplying rational expressions in lowest terms in your MyMathLab course.

Divide. Write each answer in lowest terms.

16. $\dfrac{4m-12}{2m+10} \div \dfrac{9-m^2}{m^2-25}$

16. _____

Copyright © 2015 Pearson Education, Inc.

CHAPTER 1 Equations and Inequalities

Name: Date:
Instructor: Section:

17. $\dfrac{27-3k^2}{3k^2+8k-3} \div \dfrac{k^2-6k+9}{6k^2-19k+3}$ 17. _____

18. $\dfrac{y^2+7y+10}{3y+6} \div \dfrac{y^2+2y-15}{4y-4}$ 18. _____

For extra help for exercise 19, see the videos on solving equations that are quadratic in form in your MyMathLab course.

Solve each equation. Check your solutions.

19. $4t^4 = 21t^2 - 5$ 19. _____

For extra help for exercises 20–22, see the videos on simplifying expressions with negative exponents in your MyMathLab course.

Evaluate or simplify each expression, and write it using only positive exponents. Assume that all variables represent nonzero real numbers.

20. $-2k^{-4}$ 20. _____

21. $(m^2n)^{-9}$ 21. _____

Name: Date:
Instructor: Section:

22. $\dfrac{2x^{-4}}{3y^{-7}}$

22. _____

For extra help for exercises 23–28, see the videos on evaluating exponentials with rational exponents in your MyMathLab course.

Evaluate each exponential.

23. $-256^{1/4}$

23. _____

24. $16^{1/2}$

24. _____

25. $(-3375)^{1/3}$

25. _____

26. $-81^{5/4}$

26. _____

27. $36^{5/2}$

27. _____

28. $\left(\dfrac{125}{27}\right)^{-2/3}$

28. _____

For extra help for exercises 29–31, see the videos on writing exponential expressions as radicals in your MyMathLab course.

Write with radicals. Assume that all variables represent positive real numbers.

29. $4y^{2/5} + (5x)^{1/5}$

29. _____

CHAPTER 1 Equations and Inequalities

Name: Date:
Instructor: Section:

30. $\left(2x^4 - 3y^2\right)^{-4/3}$

30. _____

Simplify the radical by rewriting it with a rational exponent. Write answers in radical form if necessary. Assume that variables represent positive real numbers.

31. $\sqrt[8]{a^2}$

31. _____

For extra help for exercises 32–34, see the videos on evaluating exponentials with rational exponents in your MyMathLab course.

Use the rules of exponents to simplify each expression. Write all answers with positive exponents. Assume that variables represent positive real numbers.

32. $y^{7/3} \cdot y^{-4/3}$

32. _____

33. $\dfrac{a^{2/3} \cdot a^{-1/3}}{\left(a^{-1/6}\right)^3}$

33. _____

34. $\dfrac{\left(x^{-3}y^2\right)^{2/3}}{\left(x^2 y^{-5}\right)^{2/5}}$

34. _____

For extra help for exercises 35–37, see the videos on finding cube, fourth, and other roots in your MyMathLab course.

Find each root.

35. $\sqrt[3]{-64}$

35. _____

36. $\sqrt[4]{256}$

36. _____

Name: Date:
Instructor: Section:

37. $\sqrt[7]{-1}$

37. _____

For extra help for exercises 38–40, see the videos on finding *n*th roots of *n*th powers in your MyMathLab course.

Simplify each root.

38. $\sqrt{(-9)^2}$

38. _____

39. $-\sqrt[5]{x^5}$

39. _____

40. $-\sqrt[4]{x^{16}}$

40. _____

For extra help for exercises 41–43, see the videos on simplifying radicals in your MyMathLab course.

Simplify each radical. Assume that variables represent positive real numbers.

41. $\sqrt[42]{x^{28}}$

41. _____

42. $\sqrt[3]{1250a^5b^7}$

42. _____

43. $\sqrt[3]{\dfrac{5}{49x}}$

43. _____

For extra help for exercises 44–47, see the videos on simplifying products and quotients of radicals with different indexes in your MyMathLab course.

Simplify each radical. Assume that variables represent positive real numbers.

44. $\sqrt{r} \cdot \sqrt[3]{r}$

44. _____

CHAPTER 1 Equations and Inequalities

Name: Date:
Instructor: Section:

45. $\sqrt[4]{2}\cdot\sqrt[8]{7}$ 45. _____

46. $\sqrt{3}\cdot\sqrt[5]{64}$ 46. _____

47. $\left(2+\sqrt[3]{5}\right)\left(2-\sqrt[3]{5}\right)$ 47. _____

Name: Date:
Instructor: Section:

Chapter 1 Equations and Inequalities

1.7R   Order on the Number Line; Sets and Set Operations

Key Terms

Use the vocabulary terms listed below to complete each statement in exercises 1–5.

    intersection   compound inequality   union

    negative number   positive number

1. A number located to the left of 0 on a number line is a _____.

2. A number located to the right of 0 on a number line is a _____.

3. The _____ of two sets, A and B, is the set of elements that belong to either A or B or both.

4. A _____ is formed by joining two inequalities with a connective word such as *and* or *or*.

5. The _____ of two sets, A and B, is the set of elements that belong to both A or B.

Guided Examples

Review these examples:

1. Is the statement $-4 < -2$ true or false?

   Because -4 is to the left of -2 on the number line, -4 is less than -2. The statement $-4 < -2$ is true.

   $-5\;-4\;-3\;-2\;-1\;\;0\;\;1\;\;2\;\;3\;\;4\;\;5$

2. Let $A = \{0, 1, 2, 3\}$ and $B = \{2, 3, 4, 5\}$. Find $A \cap B$.

   $A \cap B = \{2, 3\}$

Now Try:

1. Is the statement $-10 < -8$ true false.

   _____

2. Let $A = \{-6, -5, -4\}$ and $B = \{-3, -2, -1\}$. Find $A \cap$

   _____

CHAPTER 1 Equations and Inequalities

Name: Date:
Instructor: Section:

3. Solve the compound inequality $r+2<5$ and $r+3>3$, and graph the solution set.

   Solve each inequality individually.
   $r+2<5$ and $r+3>3$
   $r<3$ $r>0$
   The solution set is $(0, 3)$.

   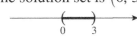

3. Solve the compound inequality $r-5\leq 3$ and $r+5\geq 3$, and graph the solution set.

   _____

   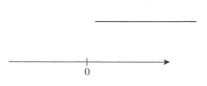

4. Solve and graph the solution set.
   $-4x-2\leq 10$ and $1-x>0$

   Solve each inequality individually.
   $-4x-2\leq 10$ and $1-x>0$
   $-4x\leq 12$ and $-x>-1$
   $x\geq -3$ and $x<1$
   The solution set is $[-3, 1)$.

   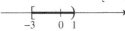

4. Solve and graph the solution set.
   $2-n>2$ and $-4n+1\leq 21$

   _____

   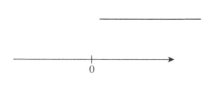

5. Solve.
   $9+x\geq 16$ and $x+7<11$

   Solve each inequality individually.
   $9+x\geq 16$ and $x+7<11$
   $x\geq 7$ and $x<4$
   The graphs of $x\geq 7$ and $x<4$ are shown below.

   There is no number that is both greater than or equal to 7 and less than 4, so the compound inequality has no solution.
   The solution set is ∅.

5. Solve.
   $6x>-36$ and $3x\leq -24$

   _____

6. Let $A=\{0, 1, 2, 3\}$ and $B=\{2, 3, 4, 5\}$. Find $A\cup B$.

   $A\cup B=\{0, 1, 2, 3, 4, 5\}$

6. Let $A=\{-6,-5,-4\}$ and $B=\{-3,-2,-1\}$. Find $A\cup B$.

   _____

Name: Date:
Instructor: Section:

7. Solve the compound inequality $m+1>5$ or $-2m>2$ and graph the solution set.

   Solve each inequality individually.
   $m+1>5$ or $-2m>2$
   $m>4$ or $m<-1$
   The solution set is $(-\infty,-1)\cup(4,\infty)$.
   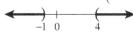

7. Solve the compound inequality $x+1\geq 3$ or $6+x<4$ and graph the solution set.

   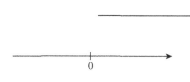

8. Solve the compound inequality $2x+1\leq 7$ or $3x-4\leq 2$ and graph the solution set.

   Solve each inequality individually.
   $2x+1\leq 7$ or $3x-4\leq 2$
   $2x\leq 6$ or $3x\leq 6$
   $x\leq 3$ or $x\leq 2$
   The graphs of $x\leq 3$ and $x\leq 2$ are shown below.

   By taking the union, we find that the solution set is $(-\infty, 3]$.
   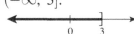

8. Solve the compound inequality $x-4\geq 3$ or $2x>18$ and graph the solution set.

   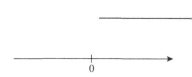

IRW-106 CHAPTER 1 Equations and Inequalities

Name: Date:
Instructor: Section:

9. Solve the compound inequality $2-n<8$ or $-4n+1 \geq 21$ and graph the solution set.

   Solve each inequality individually.
   $2-n<8$ or $-4n+1 \geq 21$
   $-n<6$ or $-4n \geq 20$
   $n>-6$ or $n \leq -5$

   The graphs of $n>-6$ and $n \leq -5$ are shown below.

   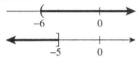

   By taking the union, we obtain every real number as a solution, since every real number satisfies at least one of the two inequalities. The solution set is $(-\infty, \infty)$.

   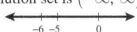

9. Solve the compound inequality $4n-6 \leq 6n$ or $-n > 4n-10$ and graph the solution set.

   9. _____

   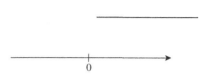

Practice Problems

For extra help for exercises 1–3, see the videos on order on the number line in your MyMathLab course.

Decide whether each statement is **true** *or* **false**.

1. $-76 < 45$ 1. _____

2. $-5 > -5$ 2. _____

3. $-12 > -10$ 3. _____

For extra help for exercises 4–6, see the videos on finding the intersection of two sets in your MyMathLab course.

Let $A = \{0, 1, 2, 3, 4, 5\}$, $B = \{2, 4, 6, 8, 10\}$, $C = \{1, 3, 5, 7, 9\}$, $D = \{0, 2, 4\}$, *and* $E = \{0\}$. *Specify each set.*

4. $A \cap C$ 4. _____

5. $B \cap D$ 5. _____

Name: Date:
Instructor: Section:

6. $C \cap E$ 6. _____

For extra help for exercises 7–9, see the videos on solving compound inequalities with the word *and* in your MyMathLab course.

For each compound inequality, give the solution set in both interval and graph forms.

7. $x - 3 \leq 6$ and $x + 2 \geq 7$ 7. _____

<-+-+-+-+-+-+-+-+-+-+-+->

8. $m - 7 \leq -3$ and $m + 2 < -3$ 8. _____

<-+-+-+-+-+-+-+-+-+-+-+->

9. $5t > 0$ and $5t + 4 \leq -1$ 9. _____

<-+-+-+-+-+-+-+-+-+-+-+->

For extra help for exercises 10–12, see the videos on finding the union of two sets in your MyMathLab course.

Let $A = \{0, 1, 2, 3, 4, 5\}$, $B = \{2, 4, 6, 8, 10\}$, $C = \{1, 3, 5, 7, 9\}$, $D = \{1, 2, 3\}$, and $E = \{0\}$. *Specify each set.*

10. $A \cup D$ 10. _____

11. $B \cup C$ 11. _____

12. $C \cup E$ 12. _____

Copyright © 2015 Pearson Education, Inc.

Name: Date:
Instructor: Section:

For extra help for exercises 13–14, see the videos on solving compound inequalities with the word *or* in your MyMathLab course.

For each compound inequality, give the solution set in both interval and graph forms.

13. $q + 3 > 7$ or $q + 1 \leq -3$ 13. _____

14. $3 > 4m + 2$ or $7m - 3 \geq -2$ 14. _____

Chapter 1 Equations and Inequalities

1.8R Definition and Properties of Absolute Value; Evaluating Absolute Value Expressions

Key Terms

Use the vocabulary terms listed below to complete each statement in exercise 1.

absolute value

1. The _____ of a number is the distance between 0 and the number on the number line.

Guided Examples

Review these examples:

1. Simplify by finding the absolute value.

    a. $|0|$

    $|0| = 0$

    b. $|16|$

    $|16| = 16$

    c. $|-16|$

    $|-16| = -(-16) = 16$

    d. $-|16|$

    $-|16| = -(16) = -16$

    e. $-|-16|$

    $-|-16| = -(16) = -16$

    f. $|7-5|$

    $|7-5| = |2| = 2$

Now Try:

1. Simplify by finding the absolute value.

    a. $|10|$

    b. $|-10|$

    c. $-|10|$

    d. $-|-10|$

    e. $|10-7|$

    f. $-|10-7|$

CHAPTER 1 Equations and Inequalities

Name: Date:
Instructor: Section:

g. $-|7-5|$

$-|7-5| = -|2| = -2$

g. $-|7-10|$

_____

Practice Problems

For extra help for exercises 1–3, see the videos on finding the absolute value of a real number in your MyMathLab course.

Simplify.

1. $-|49-39|$ 1. _____

2. $|-7.52+6.3|$ 2. _____

3. $|16-14|$ 3. _____

Copyright © 2015 Pearson Education, Inc.

Name: Date:
Instructor: Section:

Chapter 2 Graphs and Functions

2.1R   Evaluating Expressions for Given Values

Key Terms

Use the vocabulary terms listed below to complete each statement in exercises 1–3.

variable   constant   algebraic expression

1. A _____ is a symbol, usually a letter, used to represent an unknown number.

2. A collection of numbers, variables, operation symbols, and grouping symbols is an_____.

3. A _____ is a fixed, unchanging number.

Guided Examples

Review these examples:

1. Find the value of each algebraic expression for $p = 4$ and then $p = 7$.

   a. $6p$

   For $p = 4$,
   $6p = 6 \cdot 4$ Let $p = 4$.
   $ = 24$ Multiply.

   For $p = 7$,
   $6p = 6 \cdot 7$ Let $p = 7$.
   $ = 42$ Multiply.

   b. $5p^2$

   For $p = 4$,
   $5p^2 = 5 \cdot 4^2$ Let $p = 4$.
   $ = 5 \cdot 16$ Square 4.
   $ = 80$ Multiply.

   For $p = 7$,
   $5p^2 = 5 \cdot 7^2$ Let $p = 7$.
   $ = 5 \cdot 49$ Square 7.
   $ = 245$ Multiply.

Now Try:

1. Find the value of each algebraic expression for $k = 6$ and then $k = 9$.

   a. $4k$

   _____

   b. $7k^2$

   _____

CHAPTER 2 Graphs and Functions

Name: Date:
Instructor: Section:

2. Find the value of each expression for $x = 7$ and $y = 6$.

   a. $3x + 4y$

   Replace x with 7 and y with 6.
   $3x + 4y = 3 \cdot 7 + 4 \cdot 6$
   $ = 21 + 24$ Multiply.
   $ = 45$ Add.

   b. $\dfrac{8x - 6y}{4x - 3y}$

   Replace x with 7 and y with 6.
   $\dfrac{8x - 6y}{4x - 3y} = \dfrac{8 \cdot 7 - 6 \cdot 6}{4 \cdot 7 - 3 \cdot 6}$

   $\phantom{\dfrac{8x - 6y}{4x - 3y}} = \dfrac{56 - 36}{28 - 18}$ Multiply.

   $\phantom{\dfrac{8x - 6y}{4x - 3y}} = \dfrac{20}{10}$ Subtract.

   $\phantom{\dfrac{8x - 6y}{4x - 3y}} = 2$ Divide.

   c. $3x^2 - 4y^2$

   Replace x with 7 and y with 6.
   $3x^2 - 4y^2$
   $= 3 \cdot 7^2 - 4 \cdot 6^2$
   $= 3 \cdot 49 - 4 \cdot 36$ Apply the exponents.
   $= 147 - 144$ Multiply.
   $= 3$ Subtract.

2. Find the value of each expression for $x = 8$ and $y = 4$.

   a. $5x + 6y$

   _____

   b. $\dfrac{9x + 2y}{3x - 5y}$

   _____

   c. $2x^2 - 3y^2$

   _____

Practice Problems

For extra help for exercises 1–3, see the videos on evaluating algebraic expressions given values for the variables in your MyMathLab course.

Find the value of each expression if $x = 2$ and $y = 4$.

1. $9x - 3y + 2$

1. _____

Copyright © 2015 Pearson Education, Inc.

Name: Date:
Instructor: Section:

2. $\dfrac{2x+3y}{3x-y+2}$

2. _____

3. $\dfrac{3y^2+2x^2}{5x+y^2}$

3. _____

Name: Date:
Instructor: Section:

Chapter 2 Graphs and Functions

2.2R   Squaring a Binomial; Factoring Perfect Square Trinomials

For review material about factoring perfect square trinomials, see section 1.4/1.5R Guided Examples 14 and 15 and Practice Problems 13–15.

Key Terms
Use the vocabulary terms listed below to complete each statement in exercises 1–3.

    binomial   **perfect square trinomial**

1. A polynomial with two terms is called a _____.

2. A _____ is a trinomial that can be factored as the square of a binomial.

Guided Examples

Review these examples: | **Now Try:**

1. Find $(m+5)^2$.

    $(m+5)^2 = (m+5)(m+5)$
    $= m^2 + 5m + 5m + 25$
    $= m^2 + 10m + 25$

1. Find $(x+6)^2$.

    _____

2. Square each binomial.

    a. $(7z-4)^2$

    $(7z-4)^2 = (7z)^2 - 2(7z)(4) + (-4)^2$
    $= 7^2 z^2 - 56z + 16$
    $= 49z^2 - 56z + 16$

    b. $(6x+3y)^2$

    $(6x+3y)^2 = (6x)^2 + 2(6x)(3y) + (3y)^2$
    $= 36x^2 + 36xy + 9y^2$

    c. $(5a-7x)^2$

    $(5a-7x)^2 = (5a)^2 - 2(5a)(7x) + (7x)^2$
    $= 25a^2 - 70ax + 49x^2$

2. Square each binomial.

    a. $(8a-3b)^2$

    _____

    b. $(2a+9k)^2$

    _____

    c. $(3a-11b)^2$

    _____

Copyright © 2015 Pearson Education, Inc.

Name: Date:
Instructor: Section:

d. $\left(6n+\dfrac{1}{4}\right)^2$

$\left(6n+\dfrac{1}{4}\right)^2 = (6n)^2 + 2(6n)\left(\dfrac{1}{4}\right) + \left(\dfrac{1}{4}\right)^2$

$= 36n^2 + 3n + \dfrac{1}{16}$

e. $2x(5x-4)^2$

$2x(5x-4)^2 = 2x(25x^2 - 40x + 16)$

$= 50x^3 - 80x^2 + 32x$

d. $\left(3p+\dfrac{1}{6}\right)^2$

_____

e. $5x(3x-7)^2$

_____

Practice Problems

For extra help for exercises 1–3, see the videos on squaring binomials in your MyMathLab course.

Find each square by using the pattern for the square of a binomial.

1. $(7+x)^2$

1. _____

2. $(2m-3p)^2$

2. _____

3. $(4y-0.7)^2$

3. _____

Name: Date:
Instructor: Section:

Chapter 2 Graphs and Functions

2.3R Solving for a Specified Variable

Key Terms

Use the vocabulary terms listed below to complete each statement in exercise 1.

formula

1. An equation in which variables are used to describe a relationship is called a(n) _____.

Guided Examples

Review these examples:

1. Find the value of the remaining variable in each formula.

   a. $A = LW$; $A = 54, L = 8$

   Substitute the given values for A and L into the formula.

   $A = LW$

   $54 = 8W$

   $\dfrac{54}{8} = \dfrac{8W}{8}$

   $6.75 = W$

   The width is 6.75. Since $8(6.75) = 54$, the answer checks.

Now Try:

1. Find the value of the remaining variable in each formula.

   a. $A = LW$; $A = 88, L = 16$

   _____

Name: Date:
Instructor: Section:

b. $A = \frac{1}{2}h(b+B)$; $A = 216$, $B = 21$, $h = 12$

Substitute the given values for A, B, and h into the formula.

$$A = \frac{1}{2}h(b+B)$$
$$216 = \frac{1}{2}(12)(b+21)$$
$$216 = 6(b+21)$$
$$216 = 6b+126$$
$$216 - 126 = 6b+126-126$$
$$90 = 6b$$
$$\frac{90}{6} = \frac{6b}{6}$$
$$15 = b$$

The length of the shorter side b is 15. Since $\frac{1}{2}(12)(15+21) = 216$, the answer checks.

2. Solve $A = \frac{1}{2}bh$ for h.

$$A = \frac{1}{2}bh$$
$$2A = bh$$
$$\frac{2A}{b} = \frac{bh}{b}$$
$$\frac{2A}{b} = h \quad \text{or} \quad h = \frac{2A}{b}$$

3. Solve $A = p + prt$ for r.

$$A = p + prt$$
$$A - p = p + prt - p$$
$$A - p = prt$$
$$\frac{A-p}{pt} = \frac{prt}{pt}$$
$$\frac{A-p}{pt} = r \quad \text{or} \quad r = \frac{A-p}{pt}$$

b. $A = \frac{1}{2}h(b+B)$; $A = 740$, $h = 20$, $B = 43$

_____

2. Solve $d = rt$ for t.

_____

3. Solve $P = a + b + c$ for a.

_____

Name: Date:
Instructor: Section:

4. Solve $V = k + gt$ for t.

$$V = k + gt$$
$$V - k = k + gt - k$$
$$V - k = gt$$
$$\frac{V-k}{g} = \frac{gt}{g}$$
$$\frac{V-k}{g} = t$$
$$t = \frac{V-k}{g}$$

4. Solve $A = \frac{1}{2}h(b+B)$ for h.

_____

5. Solve each equation for y.

a. $5x - y = 9$

$$5x - y = 9$$
$$5x - y - 5x = 9 - 5x$$
$$-y = 9 - 5x$$
$$-1(-y) = -1(9 - 5x)$$
$$y = -9 + 5x \quad \text{or} \quad y = 5x - 9$$

b. $-4x + 5y = 15$

$$-4x + 5y = 15$$
$$-4x + 5y + 4x = 15 + 4x$$
$$5y = 4x + 15$$
$$\frac{5y}{5} = \frac{4x+15}{5}$$
$$y = \frac{4x}{5} + \frac{15}{5}$$
$$y = \frac{4}{5}x + 3$$

5. Solve each equation for y.

a. $6x - y = 18$

_____

b. $-18x + 3y = 15$

_____

Name:
Instructor:

Date:
Section:

Practice Problems

For extra help for exercises 1–3, see the videos on solving a formula for one variable, given the values of the other variables, in your MyMathLab course.

In the following exercises, a formula is given, along with the values of all but one of the variables in the formula. Find the value of the variable that is not given.

1. $S = \dfrac{a}{1-r}$; $S = 60$, $r = .4$

1. _____

2. $I = prt$; $I = 288$, $r = .04$, $t = 3$

2. _____

3. $A = \frac{1}{2}(b+B)h$; $b = 6$, $B = 16$, $A = 132$

3. _____

For extra help for exercises 4–6, see the videos on solving a formula for a specified variable in your MyMathLab course.

Solve each formula for the specified variable.

4. $V = LWH$ for H

4. _____

5. $S = (n-2)180$ for n

5. _____

6. $V = \frac{1}{3}\pi r^2 h$ for h

6. _____

Copyright © 2015 Pearson Education, Inc.

Chapter 2 Graphs and Functions

2.4R   Division Involving Zero

Key Terms

Use the vocabulary terms listed below to complete each statement in exercise 1.

    dividend   divisor   quotient   remainder

1. The number left over when two numbers do not divide exactly is the _____.

2. The number being divided by another number in a division problem is the _____.

3. The answer to a division problem is called the _____.

4. In the problem $639 \div 9$, 9 is called the _____.

Guided Examples

Review these examples:

1. Divide.

   a. $0 \div 17$

   $0 \div 17 = 0$

   b. $0 \div 9674$

   $0 \div 9674 = 0$

   c. $\dfrac{0}{668}$

   $\dfrac{0}{668} = 0$

   d. $139\overline{)0}$

   $139\overline{)\,0\,}^{\,0}$

2. Divide.

   a. $\dfrac{13}{0}$

   $\dfrac{13}{0}$ is undefined.

Now Try:

1. Divide.

   a. $0 \div 6$

   _____

   b. $0 \div 1283$

   _____

   c. $\dfrac{0}{25}$

   _____

   d. $67\overline{)0}$

   _____

2. Divide.

   a. $\dfrac{9}{0}$

   _____

Name: Date:
Instructor: Section:

    b. $0\overline{)21}$ **b.** $0\overline{)21}$

       $0\overline{)21}$ is undefined.   _____

    c. $27 \div 0$ **c.** $110 \div 0$

       $27 \div 0$ is undefined.   _____

For extra help for exercises 1–6, see the videos on division involving 0 in your MyMathLab course.

*Divide. If the division is not possible, write "**undefined**."*

    1. $12\overline{)0}$ **1.** _____

    2. $\dfrac{0}{6}$ **2.** _____

    3. $0 \div 15$ **3.** _____

    4. $0\overline{)72}$ **4.** _____

    5. $\dfrac{7}{0}$ **5.** _____

    6. $9 \div 0$ **6.** _____

Chapter 2 Graphs and Functions

2.8R Operations with Polynomials; Operations with Rational Expressions

For review material about adding, subtracting, and multiplying polynomials, see section 1.3 Guided Examples 16–27 and Practice Problems 22–39.

For review material about operations with rational expressions, see section 1.6 Guided Examples 1–12 and Practice Problems 1–18.

For review material about dividing a polynomial by a polynomial, see section 3.2R Guided Examples 1–5 and Practice Problems 1–3.

Key Terms

Use the vocabulary terms listed below to complete each statement in exercises 1–3.

      quotient   dividend   divisor

1. In the division $\dfrac{5x^5 - 10x^3}{5x^2} = x^3 - 2x$, the expression $5x^5 - 10x^3$ is the

   _____.

2. In the division $\dfrac{5x^5 - 10x^3}{5x^2} = x^3 - 2x$, the expression $x^3 - 2x$ is the _____.

3. In the division $\dfrac{5x^5 - 10x^3}{5x^2} = x^3 - 2x$, the expression $5x^2$ is the _____.

Guided Examples

Review these examples:

1. Divide $6x^4 - 18x^3$ by $6x^2$.

   $$\dfrac{6x^4 - 18x^3}{6x^2} = \dfrac{6x^4}{6x^2} - \dfrac{18x^3}{6x^2}$$
   $$= x^2 - 3x$$

   Check Multiply. $6x^2(x^2 - 3x) = 6x^4 - 18x^3$

Now Try:

1. Divide $6x^4 - 18x^3$ by $6x^2$.

   _____

Name: Date:
Instructor: Section:

2. Divide. $\dfrac{25a^6 - 15a^4 + 10a^2}{5a^3}$

   Divide each term by $5a^3$.
   $$\dfrac{25a^6 - 15a^4 + 10a^2}{5a^3} = \dfrac{25a^6}{5a^3} - \dfrac{15a^4}{5a^3} + \dfrac{10a^2}{5a^3}$$
   $$= 5a^3 - 3a + \dfrac{2}{a}$$

   Check:
   $$5a^3\left(5a^3 - 3a + \dfrac{2}{a}\right)$$
   $$= 5a^3(5a^3) + 5a^3(-3a) + 5a^3\left(\dfrac{2}{a}\right)$$
   $$= 25a^6 - 15a^4 + 10a^2$$

3. Divide $-12x^4 + 15x^5 - 5x$ by $-5x$.

   Write the polynomial in descending powers before dividing.
   $$\dfrac{15x^5 - 12x^4 - 5x}{-5x} = \dfrac{15x^5}{-5x} - \dfrac{12x^4}{-5x} - \dfrac{5x}{-5x}$$
   $$= -3x^4 + \dfrac{12}{5}x^3 + 1$$

   Check:
   $$-5x\left(-3x^4 + \dfrac{12}{5}x^3 + 1\right)$$
   $$= -5x(-3x^4) - 5x\left(\dfrac{12}{5}x^3\right) - 5x(1)$$
   $$= 15x^5 - 12x^4 - 5x$$

2. Divide. $\dfrac{27n^5 - 36n^4 - 18}{9n^3}$

   _____

3. Divide $-8z^5 + 7z^6 - 10z$ by $2z^2$.

   _____

Practice Problems

For extra help for exercises 1–3, see the videos on dividing a polynomial by a monomial in your MyMathLab course.

Perform each division.

1. $\dfrac{16a^5 - 24a^3}{8a^2}$

1. _____

CHAPTER 2 Graphs and Functions

Name:
Instructor:

Date:
Section:

2. $\dfrac{12z^5 + 28z^4 - 8z^3 + 3z}{4z^3}$

2. _____

3. $\dfrac{39m^4 - 12m^3 + 15}{-3m^2}$

3. _____

Name: Date:
Instructor: Section:

Chapter 3 Polynomial and Rational Functions

3.1R   Squaring a Binomial; Factoring Perfect Square Trinomials

For review material about squaring a binomial, see section 2.2R Guided Examples 1 and 2 and Practice Problems 13–15.

For review material about factoring perfect square trinomials, see section 1.4/1.5R Guided Examples 14 and 15 and Practice Problems 1–3.

Chapter 3 Polynomial and Rational Functions

3.2R Dividing Polynomials

For review material about dividing a polynomial by a monomial, see section 2.8R Guided Examples 1–3 and Practice Problems 1–3.

Key Terms

Use the vocabulary terms listed below to complete each statement in exercises 1–3.

    quotient   dividend   divisor

1. In the division $\dfrac{6x^2 - 9x - 12}{2x - 5} = 3x + 3 + \dfrac{3}{2x - 5}$, the expression $2x - 5$ is the

   _____.

2. In the division $\dfrac{6x^2 - 9x - 12}{2x - 5} = 3x + 3 + \dfrac{3}{2x - 5}$, the expression $3x + 3 + \dfrac{3}{2x - 5}$ is the _____.

3. In the division $\dfrac{6x^2 - 9x - 12}{2x - 5} = 3x + 3 + \dfrac{3}{2x - 5}$, the expression $6x^2 - 9x - 12$ is the

   _____.

Guided Examples

1. Divide $225x^5y^9 - 150x^3y^7 + 110x^2y^5 - 80xy^3 + 75y^2$ by $-25xy^2$.

   $\dfrac{225x^5y^9 - 150x^3y^7 + 110x^2y^5 - 80xy^3 + 75y^2}{-25xy^2}$

   $= \dfrac{225x^5y^9}{-25xy^2} - \dfrac{150x^3y^7}{-25xy^2} + \dfrac{110x^2y^5}{-25xy^2} - \dfrac{80xy^3}{-25xy^2} + \dfrac{75y^2}{-25xy^2}$

   $= -9x^4y^7 + 6x^2y^5 - \dfrac{22xy^3}{5} + \dfrac{16y}{5} - \dfrac{3}{x}$

   Check by multiplying the quotient by the divisor.

1. Divide $80a^5b^3 + 160a^4b^2 - 120a^2b$ by $-40a^2b$.

   _____

Name:
Instructor:

Date:
Section:

2. Divide $8x + 9x^3 - 7 - 9x^2$ by $3x - 1$.

Write the dividend in descending powers as $9x^3 - 9x^2 + 8x - 7$.

Step 1 $9x^3$ divided by $3x$ is $3x^2$.
$3x^2(3x-1) = 9x^3 - 3x^2$

Step 2 Subtract. Bring down the next term.

Step 3 $-6x^2$ divided by $3x$ is $-2x$.
$-2x(3x-1) = -6x^2 + 2x$

Step 4 Subtract. Bring down the next term.

Step 5 $6x$ divided by $3x$ is 2. $2(3x-1) = 6x - 2$

$$\begin{array}{r}
3x^2 - 2x + 2 \\
3x-1\overline{\smash{)}9x^3 - 9x^2 + 8x - 7} \\
\underline{9x^3 - 3x^2} \\
-6x^2 + 8x \\
\underline{-6x^2 + 2x} \\
6x - 7 \\
\underline{6x - 2} \\
-5
\end{array}$$

$$\frac{9x^3 - 9x^2 + 8x - 7}{3x - 1} = 3x^2 - 2x + 2 + \frac{-5}{3x - 1}$$

Step 7 Multiply to check.

Check:
$(3x-1)\left(3x^2 - 2x + 2 + \frac{-5}{3x-1}\right)$

$= (3x-1)(3x^2) + (3x-1)(-2x)$

$ + (3x-1)(2) + (3x-1)\left(\frac{-5}{3x-1}\right)$

$= 9x^3 - 3x^2 - 6x^2 + 2x + 6x - 2 - 5$

$= 9x^3 - 9x^2 + 8x - 7$

2. Divide $-12x^2 + 10x^3 - 3 - 8x$ by $5x - 1$.

_____

CHAPTER 3 Polynomial and Rational Functions

Name: Date:
Instructor: Section:

3. Divide $x^3 - 64$ by $x - 4$.

   Here the dividend is missing the x^2-term and the x-term. We use 0 as the coefficient for each missing term.

   $$\begin{array}{r} x^2 + 4x + 16 \\ x-4 \overline{\smash{)}x^3 + 0x^2 + 0x - 64} \\ \underline{x^3 - 4x^2} \\ 4x^2 + 0x \\ \underline{4x^2 - 16x} \\ 16x - 64 \\ \underline{16x - 64} \\ 0 \end{array}$$

   The remainder is 0. The quotient is $x^2 + 4x + 16$.
   Check:
   $(x-4)(x^2 + 4x + 16)$
   $= x^3 + 4x^2 + 16x - 4x^2 - 16x - 64$
   $= x^3 - 64$

4. Divide $x^4 - 3x^3 + 7x^2 - 8x + 14$ by $x^2 + 2$.

   Since $x^2 + 2$ is missing the x-term, we write it as $x^2 + 0x + 2$.

   $$\begin{array}{r} x^2 - 3x + 5 \\ x^2+0x+2 \overline{\smash{)}x^4 - 3x^3 + 7x^2 - 8x + 14} \\ \underline{x^4 + 0x^3 + 2x^2} \\ -3x^3 + 5x^2 - 8x \\ \underline{-3x^3 + 0x^2 - 6x} \\ 5x^2 - 2x + 14 \\ \underline{5x^2 + 0x + 10} \\ -2x + 4 \end{array}$$

   The quotient is $x^2 - 3x + 5 + \dfrac{-2x+4}{x^2+2}$

   The check shows that the quotient multiplied by the divisor gives the original dividend.

3. Divide $x^3 - 1000$ by $x - 10$.

   _____

4. Divide
   $3x^4 + 5x^3 - 7x^2 - 12x + 9$
   by $x^2 - 4$.

   _____

Name: Date:
Instructor: Section:

5. Divide $5x^3 + 8x^2 + 12x - 1$ by $5x + 5$.

$$\begin{array}{r} x^2 + \dfrac{3}{5}x + \dfrac{9}{5} \\ 5x+5{\overline{\smash{\big)}\,5x^3 + 8x^2 + 12x - 1}} \\ \underline{5x^3 + 5x^2} \\ 3x^2 + 12x \\ \underline{3x^2 + 3x} \\ 9x - 1 \\ \underline{9x + 9} \\ -10 \end{array}$$

The answer is $x^2 + \dfrac{3}{5}x + \dfrac{9}{5} - \dfrac{10}{5x+5}$.

5. Divide $8x^3 - 7x^2 + 4x + 1$ by $8x - 8$.

_____

Practice Problems

For extra help for exercises 1–3, see the videos on dividing a polynomial by a polynomial in your MyMathLab course.

Perform each division.

1. $\dfrac{-6x^2 + 23x - 20}{2x - 5}$

1. _____

Copyright © 2015 Pearson Education, Inc.

CHAPTER 3 Polynomial and Rational Functions

Name: Date:
Instructor: Section:

2. $\dfrac{6x^4 - 12x^3 + 13x^2 - 5x - 1}{2x^2 + 3}$ 2. _____

3. $\dfrac{2a^4 + 5a^2 + 3}{2a^2 + 3}$ 3. _____

Name: Date:
Instructor: Section:

Chapter 3 Polynomial and Rational Functions

3.6R   Using Geometric Formulas; Ratio and Proportion

Key Terms

Use the vocabulary terms listed below to complete each statement in exercises 1–8.

formula	**area**	**perimeter**
vertical angles	**denominator**	**numerator**
ratio	**cross products**	**proportion**

1. The nonadjacent angles formed by two intersecting lines are called _____.

2. An equation in which variables are used to describe a relationship is called a(n) _____.

3. The distance around a figure is called its _____.

4. A measure of the surface covered by a figure is called its _____.

5. A _____ can be used to compare two measurements with the same type of units.

6. When writing the ratio to compare the width of a room to its height, the width goes in the _____ and the height goes in the _____.

7. A _____ shows that two ratios or rates are equivalent.

8. To see whether a proportion is true, determine if the _____ are equal.

Copyright © 2015 Pearson Education, Inc.

Guided Examples

Review these examples:

1. Find the dimensions of a rectangle. The length is 4 m less than three times the width. The perimeter is 96 m.

   Step 1 Read the problem. We must find the dimensions of the rectangle.

   Step 2 Assign a variable.
   Let W = the width of the rectangle, in meters. Then $L = 3W - 4$ is the length, in meters.

   Step 3 Write an equation. Use the formula for the perimeter of a rectangle. Substitute $3W - 4$ for the length.
   $$P = 2L + 2W$$
   $$96 = 2(3W - 4) + 2W$$

   Step 4 Solve.
   $$96 = 6W - 8 + 2W$$
   $$96 = 8W - 8$$
   $$96 + 8 = 8W - 8 + 8$$
   $$104 = 8W$$
   $$\frac{104}{8} = \frac{8W}{8}$$
   $$13 = W$$

   Step 5 State the answer. The width is 13 m. The length is $3(13) - 4 = 35$ m.

   Step 6 Check. The perimeter is $2(13) + 2(35) = 96$ m. The answer checks.

Now Try:

1. Ruth has 42 feet of binding for a rectangular rug that she is weaving. If the rug is 9 feet wide, how long can she make the rug if she wishes to use all the binding on the perimeter of the rug?

   _____

Name: Date:
Instructor: Section:

2. The longest side of a triangle is 4 feet longer than the shortest side. The medium side is 2 feet longer than the shortest side. If the perimeter is 36 feet, what are the lengths of the three sides?

Step 1 Read the problem. We must find the lengths of the sides.

Step 2 Assign a variable.
Let s = the length of the shortest side, in feet.
Then $s + 2$ = the length of the medium side, in feet,
and $s + 4$ = the length of the longest side, in feet.

Step 3 Write an equation. Use the formula for the perimeter of a triangle.
$$P = a + b + c$$
$$36 = s + (s+2) + (s+4)$$

Step 4 Solve.
$$36 = 3s + 6$$
$$30 = 3s$$
$$10 = s$$

Step 5 State the answer. Since s represents the length of the shortest side, its measure is 10 ft.
$s + 2 = 10 + 2 = 12$ ft is the length of the medium side.
$s + 4 = 10 + 4 = 14$ ft is the length of the longest side.

Step 6 Check. The perimeter is
$10 + 12 + 14 = 36$ ft, as required.

2. The longest side of a triangle is twice as long as the shortest side. The medium side is 5 feet longer than the shortest side. If the perimeter is 65 feet, what are the lengths of the three sides?

_____

Name: Date:
Instructor: Section:

3. The area of a triangle is 104 ft². The base of the triangle is 16 ft. Find the height of the triangle.

   Step 1 Read the problem. We must find the height of the triangle

   Step 2 Assign a variable.
   Let h = the height of the triangle, in feet.

   Step 3 Write an equation. Use the formula for the area of a triangle. Substitute $A = 104$ and $b = 16$.
   $$A = \frac{1}{2}bh$$
   $$104 = \frac{1}{2}(16)h$$

   Step 4 Solve.
   $$104 = 8h$$
   $$13 = h$$

   Step 5 State the answer. The height is 13 ft.

   Step 6 Check to see that the values $A = 104$, $b = 16$ and $h = 13$ satisfy the formula for the area of a triangle.

3. The area of a triangle is 275 m². The height of the triangle is 22 m. Find the base of the triangle.

   _____

4. a. Find the measure of the marked angles in the figure below.

   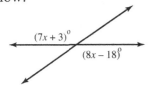

   Since the marked angles are vertical angles, they have equal measures.
   $$7x + 3 = 8x - 18$$
   $$3 = x - 18$$
   $$21 = x$$
   Since $x = 21$, one angle has measure $7(21) + 3 = 150$ degrees. The other has the same measure, since $8(21) - 18 = 150$ as well. Each angle measures 150°.

4. a. Find the measure of the marked angles in the figure below.

   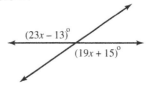

   _____

Copyright © 2015 Pearson Education, Inc.

Name: Date:
Instructor: Section:

b. Find the measure of the marked angles in the figure below.

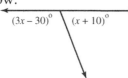

The measures of the marked angles must add to 180° because together they form a straight angle. The angles are supplements of each other.

$$(3x - 30) + (x + 10) = 180$$
$$4x - 20 = 180$$
$$4x = 200$$
$$x = 50$$

Replace x with 50 in the measure of each marked angle.

$$3x - 30 = 3(50) - 30 = 150 - 30 = 120$$
$$x + 10 = 50 + 10 = 60$$

The two angles measure 120° and 60°.

b. Find the measure of the marked angles in the figure below.

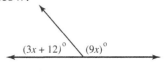

_____

5. A meeting room measures 22 feet long, 17 feet wide and 13 feet high. Write each ratio as a fraction, using the room measurements.

a. Ratio of length to width

The ratio of length to width is $\dfrac{22 \text{ ft}}{17 \text{ ft}} = \dfrac{22}{17}$

b. Ratio of width to height.

The ratio of width to height is $\dfrac{17 \text{ ft}}{13 \text{ ft}} = \dfrac{17}{13}$

5. Jacob spent $5 for breakfast, $8 for lunch, and $17 for dinner. Write each ratio as a fraction.

a. Ratio of the amount spent on a dinner to the amount spent on lunch

_____

b. Ratio of the amount spent on lunch to the amount spent on breakfast

_____

Copyright © 2015 Pearson Education, Inc.

Name: _____ Date: _____
Instructor: _____ Section: _____

6. Write each ratio in lowest terms.

a. 32 inches of snow to 8 inches of rain

The ratio is $\frac{32 \text{ inches}}{8 \text{ inches}}$. Divide out the common units. Then write this ratio in lowest terms by dividing the numerator and denominator by 8.

$$\frac{32 \text{ inches}}{8 \text{ inches}} = \frac{32}{8} = \frac{32 \div 8}{8 \div 8} = \frac{4}{1}$$

So, the ratio of 32 inches of snow to 8 inches of rain is 4 to 1, or $\frac{4}{1}$. For every 4 inches of snow there is 1 inch of rain.

b. 12 ounces of cleaner to 40 ounces of water

The ratio is $\frac{12 \text{ ounces}}{40 \text{ ounces}}$. Divide out the common units. Then write this ratio in lowest terms by dividing the numerator and denominator by 4.

$$\frac{12 \text{ ounces}}{40 \text{ ounces}} = \frac{12}{40} = \frac{12 \div 4}{40 \div 4} = \frac{3}{10}$$

So, the ratio of 12 ounces of cleaner to 40 ounces of water is $\frac{3}{10}$. For every 3 ounces of cleaner, there are 10 ounces of water.

c. 14 people at a large table to 8 people at a small table

The ratio is $\frac{14}{8} = \frac{14 \div 2}{8 \div 2} = \frac{7}{4}$

6. Write each ratio in lowest terms.

a. 16 seconds to 24 seconds

_____

b. 150 grams to 75 grams

_____

c. width of 14 m to length of 21 m

_____

Name: Date:
Instructor: Section:

7. The price of a newspaper increased from $0.75 to $1.00. Find the ratio of the increase in price to the original price.

new price − original price = increase
$1.00 − $0.75 = $0.25

The words "the original price" are mentioned second, so the original price of $0.75 is in the denominator.
The ratio of increase in price to original price is shown below.

$$\frac{0.25}{0.75}$$

Now rewrite the ratio as a ratio of whole numbers. Start by multiplying both the numerator and denominator by 100.

$$\frac{0.25}{0.75} = \frac{0.25 \times 100}{0.75 \times 100} = \frac{25}{75} = \frac{25 \div 25}{75 \div 25} = \frac{1}{3}$$

8. Write each ratio as a comparison of whole numbers in lowest terms.

a. 5 hours to $5\frac{1}{3}$ hours

Divide out the common units.

$$\frac{5 \text{ hours}}{5\frac{1}{3} \text{ hours}} = \frac{5}{5\frac{1}{3}}$$

Next, write 5 as $\frac{5}{1}$ and $5\frac{1}{3}$ as the improper fraction $\frac{16}{3}$.

$$\frac{5}{5\frac{1}{3}} = \frac{\frac{5}{1}}{\frac{16}{3}}$$

Now, rewrite the problem in horizontal format, using the "÷" symbol for division. Finally, multiply by the reciprocal of the divisor.

$$\frac{\frac{5}{1}}{\frac{16}{3}} = \frac{5}{1} \div \frac{16}{3} = \frac{5}{1} \cdot \frac{3}{16} = \frac{15}{16}$$

The ratio, in lowest terms, is $\frac{15}{16}$.

7. The original price of a burger is $4.80 and the sale price is $3.00. Find the ratio of the decrease in price to the original price.

_____

8. Write each ratio as a comparison of whole numbers in lowest terms.

a. $6\frac{1}{3}$ to 3

_____

Copyright © 2015 Pearson Education, Inc.

CHAPTER 3 Polynomial and Rational Functions

Name: _____ Date: _____
Instructor: _____ Section: _____

b. $7\frac{1}{5}$ to $2\frac{1}{4}$

Write the ratio as $\dfrac{7\frac{1}{5}}{2\frac{1}{4}}$. Then write $7\frac{1}{5}$ and $2\frac{1}{4}$ as improper fractions.

$7\frac{1}{5} = \dfrac{36}{5}$ and $2\frac{1}{4} = \dfrac{9}{4}$

The ratio is shown here.

$\dfrac{7\frac{1}{5}}{2\frac{1}{4}} = \dfrac{\frac{36}{5}}{\frac{9}{4}}$

Rewrite as a division problem in horizontal format, using the "÷" symbol for division. Then multiply by the reciprocal of the divisor.

$\dfrac{36}{5} \div \dfrac{9}{4} = \dfrac{\cancel{36}^{4}}{5} \cdot \dfrac{4}{\cancel{9}_{1}} = \dfrac{16}{5}$

b. $3\frac{2}{3}$ to $2\frac{5}{6}$

_____

9. Write the ratio of length of a board 3 ft long to the length of another board that is 42 inches long.

First, express 3 ft in inches. Because 1 ft has 12 in., 3 ft is
 $3 \cdot 12$ in. $= 36$ in.
The ratio of the lengths is

$\dfrac{3 \text{ ft}}{42 \text{ in.}} = \dfrac{36 \text{ \cancel{in.}}}{42 \text{ \cancel{in.}}} = \dfrac{36}{42}$

Write the ratio in lowest terms.

$\dfrac{36}{42} = \dfrac{36 \div 6}{42 \div 6} = \dfrac{6}{7}$

The shorter board is $\dfrac{6}{7}$ the length of the longer board.

9. Write the ratio of 20 days to 4 weeks.

_____

10. Write each proportion.

   a. 7 m is to 13 m as 28 m is to 52 m

   $\dfrac{7 \text{ \cancel{m}}}{13 \text{ \cancel{m}}} = \dfrac{28 \text{ \cancel{m}}}{52 \text{ \cancel{m}}}$ so $\dfrac{7}{13} = \dfrac{28}{52}$

10. Write each proportion.

   a. 24 ft is to 17 ft as 72 ft is to 51 ft

   _____

Name: Date:
Instructor: Section:

    b. $14 is to 8 gallons as $7 is to 4 gallons

$$\frac{\$14}{8 \text{ gallons}} = \frac{\$7}{4 \text{ gallons}}$$

11. Determine whether each proportion is true or false by writing both ratios in lowest terms.

    a. $\dfrac{7}{11} = \dfrac{16}{24}$

Write each ratio in lowest terms.

$\dfrac{7}{11} \leftarrow$ Already in lowest terms $\dfrac{16 \div 8}{24 \div 8} = \dfrac{2}{3} \leftarrow$ Lowest terms

Because $\dfrac{7}{11}$ is not equivalent to $\dfrac{2}{3}$, the proportion is false.

    b. $\dfrac{9}{15} = \dfrac{21}{35}$

Write each ratio in lowest terms.

$\dfrac{9 \div 3}{15 \div 3} = \dfrac{3}{5}$ $\dfrac{21 \div 7}{35 \div 7} = \dfrac{3}{5}$

Both ratios are equivalent to $\dfrac{3}{5}$, so the proportion is true.

12. Use cross products to see whether each proportion is true or false.

    a. $\dfrac{5}{8} = \dfrac{30}{48}$

Multiply along one diagonal, then multiply along the other diagonal.

$\dfrac{5}{8} = \dfrac{30}{48}$ $\nearrow 8 \cdot 30 = 240$
                $\searrow 5 \cdot 48 = 240$

The cross products are equal, so the proportion is true.

---

    b. $10 is to 7 cans as $60 is to 42 cans

_____

11. Determine whether each proportion is true or false by writing both ratios in lowest terms.

    a. $\dfrac{36}{28} = \dfrac{24}{18}$

_____

    b. $\dfrac{4}{12} = \dfrac{9}{27}$

_____

12. Use cross products to see whether each proportion is true or false.

    a. $\dfrac{6}{17} = \dfrac{18}{51}$

_____

CHAPTER 3 Polynomial and Rational Functions

Name: Date:
Instructor: Section:

b. $\dfrac{3\tfrac{1}{5}}{4\tfrac{2}{3}} = \dfrac{7}{10}$

b. $\dfrac{3.2}{5} = \dfrac{7}{10}$

$\dfrac{3\tfrac{1}{5}}{4\tfrac{2}{3}} = \dfrac{7}{10}$

$\nearrow 4\tfrac{2}{3} \cdot 7 = \dfrac{14}{3} \cdot \dfrac{7}{1} = \dfrac{98}{3} = 32\tfrac{2}{3}$

$\searrow 3\tfrac{1}{5} \cdot 10 = \dfrac{16}{\cancel{5}} \cdot \dfrac{\cancel{10}^{2}}{1} = \dfrac{32}{1} = 32$

The cross products are unequal, so the proportion is false.

13. Find the unknown number in each proportion. Round answers to the nearest hundredth when necessary.

a. $\dfrac{14}{x} = \dfrac{21}{18}$

Recall that ratios can be rewritten in lowest terms. Write $\dfrac{21}{18}$ in lowest terms as $\dfrac{7}{6}$, which gives the proportion $\dfrac{14}{x} = \dfrac{7}{6}$.

Step 1 Find the cross product.

$\dfrac{14}{x} = \dfrac{7}{6} \quad \begin{array}{l} \nearrow x \cdot 7 \\ \searrow 14 \cdot 6 \end{array}$

Step 2 Show that cross products are equal.
$x \cdot 7 = 14 \cdot 6$
$x \cdot 7 = 84$

Step 3 Divide both sides by 7.

$\dfrac{x \cdot \cancel{7}}{\cancel{7}} = \dfrac{84}{7}$

$x = 12$

Step 4 Check in original proportion.

$\dfrac{14}{12} = \dfrac{21}{18} \quad \begin{array}{l} \nearrow 12 \cdot 21 = 252 \\ \searrow 14 \cdot 18 = 252 \end{array}$

The cross products are equal, so 12 is the correct solution.

13. Find the unknown number in each proportion. Round answers to the nearest hundredth when necessary.

a. $\dfrac{24}{x} = \dfrac{9}{12}$

_____

Copyright © 2015 Pearson Education, Inc.

Name: Date:
Instructor: Section:

b. $\dfrac{9}{13} = \dfrac{21}{x}$

Step 1 Find the cross product.
$$\dfrac{9}{13} = \dfrac{21}{x} \quad \nearrow 13 \cdot 21 = 273$$
$$\searrow 9 \cdot x$$

Step 2 Show that cross products are equal.
$9 \cdot x = 273$

Step 3 Divide both sides by 9.
$$\dfrac{\cancel{9} \cdot x}{\cancel{9}} = \dfrac{273}{9}$$
$x = 30.33$ rounded to the nearest hundredth

Step 4 Check in original proportion.
$$\dfrac{9}{13} = \dfrac{21}{30.33} \quad \nearrow 13 \cdot 21 = 273$$
$$\searrow 9 \cdot 30.33 = 272.97$$

The cross products are slightly different because of the rounded value of x. However, they are close enough to see that the problem was done correctly and that 30.33 is the approximate solution.

b. $\dfrac{2}{3} = \dfrac{x}{16}$

CHAPTER 3 Polynomial and Rational Functions

Name:
Instructor:
Date:
Section:

14. Find the unknown number in each proportion.

a. $\dfrac{3\frac{1}{4}}{8} = \dfrac{x}{12}$

$\dfrac{3\frac{1}{4}}{8} = \dfrac{x}{12}$ $\quad\nearrow 8 \cdot x$
$\quad\searrow 3\frac{1}{4} \cdot 12$

Change $3\frac{1}{4}$ to an improper fraction and write in lowest terms.

$3\frac{1}{4} \cdot 12 = \dfrac{13}{4} \cdot \dfrac{12}{1} = \dfrac{13}{\cancel{4}} \cdot \dfrac{\cancel{12}^{3}}{1} = \dfrac{39}{1} = 39$

Show that the cross products are equal.
$8 \cdot x = 39$

Divide both sides by 8.

$\dfrac{\cancel{8} \cdot x}{\cancel{8}} = \dfrac{39}{8}$

Write the solution as a mixed number in lowest terms.

$x = \dfrac{39}{8} = 4\dfrac{7}{8}$

The unknown number is $4\dfrac{7}{8}$.

Check in original proportion.

$\dfrac{3\frac{1}{4}}{8} = \dfrac{4\frac{7}{8}}{12}$ $\quad\nearrow 8 \cdot 4\dfrac{7}{8} = \dfrac{\cancel{8}}{1} \cdot \dfrac{39}{\cancel{8}} = 39$

$\quad\searrow 3\dfrac{1}{4} \cdot 12 = \dfrac{13}{\cancel{4}} \cdot \dfrac{\cancel{12}^{3}}{1} = 39$

The cross products are equal, so $4\dfrac{7}{8}$ is the correct solution.

14. Find the unknown number in each proportion.

a. $\dfrac{4\frac{1}{3}}{5} = \dfrac{x}{3}$

_____

Copyright © 2015 Pearson Education, Inc.

Name: Date:
Instructor: Section:

b. $\dfrac{3.5}{1.4} = \dfrac{4}{x}$

Show that the products are equal.
$(3.5)(x) = (1.4)(4)$
$(3.5)(x) = 5.6$
Divide both sides by 3.5.

$$\dfrac{\cancel{(3.5)} \cdot (x)}{\cancel{3.5}} = \dfrac{5.6}{3.5}$$

$$x = \dfrac{5.6}{3.5}$$

Complete the division.

$x = 1.6 \qquad 35\overline{)56.0}\ \ ^{1.6}$

So, the unknown number is 1.6. Write the solution in the original proportion and check it by finding the cross products.

$\dfrac{3.5}{1.4} = \dfrac{4}{1.6} \quad \begin{array}{l} \nearrow 1.4 \cdot 4 = 5.6 \\ \searrow 3.5 \cdot 1.6 = 5.6 \end{array}$

The cross products are equal, so 1.6 is the correct solution.

b. $\dfrac{x}{8} = \dfrac{1.2}{1.5}$

_____

Practice Problems

For extra help for exercises 1–3, see the videos on using geometric formulas in your MyMathLab course.

Use a formula to write an equation for each of the following applications; then solve the application. (Use 3.14 as an approximation for π.)

1. Find the height of a triangular banner whose area is 48 square inches and base is 12 inches.

   1. _____

2. Linda invests $5000 at 6% simple interest and earns $450. How long did Linda invest her money?

   2. _____

Copyright © 2015 Pearson Education, Inc.

IRW-144 CHAPTER 3 Polynomial and Rational Functions

Name: Date:
Instructor: Section:

3. The circumference of a circular garden is 628 feet. Find the area of the garden. (Hint: First find the radius of the garden.)

3. _____

For extra help for exercises 4–6, see the videos on vertical angles in your MyMathLab course.

Find the measure of each marked angle.

4.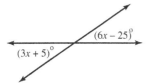

4. _____

5.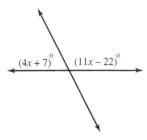

5. _____

6.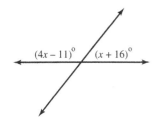

6. _____

For extra help for exercises 7–15, see the videos on writing ratios in your MyMathLab course.

Write each ratio as a fraction in lowest terms.

7. 125 cents to 95 cents

7. _____

8. 80 miles to 30 miles

8. _____

9. 5 men to 20 men

9. _____

10. $4\frac{1}{8}$ to 3

10. _____

11. 11 to $2\frac{4}{9}$

11. _____

Solve. Write each ratio as a fraction in lowest terms.

12. One car has a $15\frac{1}{2}$ gallon gas tank while another has a 22 gallon gas tank. Find the ratio of the amount the first tank holds to the amount the second tank holds.

12. _____

Write each ratio as a fraction in lowest terms. Be sure to convert units as necessary.

13. 4 days to 2 weeks

13. _____

14. 6 yards to 10 feet

14. _____

15. 40 ounces to 3 pounds

15. _____

CHAPTER 3 Polynomial and Rational Functions

Name: Date:
Instructor: Section:

For extra help for exercises 16–18, see the videos on writing proportions in your MyMathLab course.

Write each proportion.

16. 50 is to 8 as 75 is to 12.   16. _____

17. 36 is to 45 as 8 is to 10.   17. _____

18. 3 is to 33 as 12 is to 132.   18. _____

For extra help for exercises 19–24, see the videos on determining if proportions are true in your MyMathLab course.

Determine whether each proportion is true or false by writing the ratios in lowest terms. Show the simplified ratios and then write **true** *or* **false**.

19. $\dfrac{48}{36} = \dfrac{3}{4}$ 19. _____

20. $\dfrac{30}{25} = \dfrac{6}{5}$ 20. _____

21. $\dfrac{63}{18} = \dfrac{56}{14}$ 21. _____

Use cross products to determine whether each proportion is true or false. Show the cross products and then write **true** *or* **false**.

22. $\dfrac{28}{50} = \dfrac{49}{75}$ 22. _____

Name: Date:
Instructor: Section:

23. $\dfrac{4\frac{3}{5}}{9} = \dfrac{18\frac{2}{5}}{36}$ 23. _____

24. $\dfrac{2.98}{7.1} = \dfrac{1.7}{4.3}$ 24. _____

For extra help for exercises 25–30, see the videos on solving proportions in your MyMathLab course.

Find the unknown number in each proportion.

25. $\dfrac{9}{7} = \dfrac{x}{28}$ 25. _____

26. $\dfrac{7}{5} = \dfrac{98}{x}$ 26. _____

27. $\dfrac{100}{x} = \dfrac{75}{30}$ 27. _____

Find the unknown number in each proportion. Write answers as a whole or a mixed number if possible.

28. $\dfrac{2}{3\frac{1}{4}} = \dfrac{8}{x}$ 28. _____

29. $\dfrac{3}{x} = \dfrac{0.8}{5.6}$ 29. _____

30. $\dfrac{2\frac{5}{9}}{x} = \dfrac{23}{\frac{3}{5}}$ 30. _____

Copyright © 2015 Pearson Education, Inc.

IRW-148   CHAPTER 4 Inverse, Exponential, and Logarithmic Functions

Name: Date:

Instructor: Section:

Chapter 4 INVERSE, EXPONENTIAL, AND LOGARITHMIC FUNCTIONS

4.1R   Simplifying Complex Fractions

Key Terms

Use the vocabulary terms listed below to complete each statement in exercises 1–2.

complex fraction   **LCD**

1. A _____ is a rational expression with one or more fractions in the numerator, denominator, or both.

2. To simplify a complex fraction, multiply the numerator and denominator by the _____ of all the fractions within the complex fraction.

Guided Examples

Review these examples:

1. Simplify each complex fraction.

   a. $\dfrac{\dfrac{5}{6}+\dfrac{7}{12}}{\dfrac{1}{5}+\dfrac{8}{15}}$

   Step 1 Write the numerator as a single fraction.
   $$\frac{5}{6}+\frac{7}{12}=\frac{5(2)}{6(2)}+\frac{7}{12}=\frac{10}{12}+\frac{7}{12}=\frac{17}{12}$$
   Do the same with each denominator.
   $$\frac{1}{5}+\frac{8}{15}=\frac{1(3)}{5(3)}+\frac{8}{15}=\frac{3}{15}+\frac{8}{15}=\frac{11}{15}$$
   Step 2 Write the equivalent complex fraction as a division problem.
   $$\frac{\dfrac{17}{12}}{\dfrac{11}{15}}=\frac{17}{12}\div\frac{11}{15}$$
   Step 3 Divide by multiplying by the reciprocal.
   $$\frac{17}{12}\div\frac{11}{15}=\frac{17}{12}\cdot\frac{15}{11}=\frac{17\cdot5\cdot3}{3\cdot4\cdot11}=\frac{85}{44}$$

Now Try:

1. Simplify each complex fraction.

   a. $\dfrac{\dfrac{1}{6}+\dfrac{2}{9}}{\dfrac{3}{10}+\dfrac{4}{15}}$

   _____

Copyright © 2015 Pearson Education, Inc.

b. $\dfrac{8+\dfrac{4}{x}}{\dfrac{x}{6}+\dfrac{1}{12}}$ **b.** $\dfrac{9+\dfrac{3}{x}}{\dfrac{x}{10}+\dfrac{1}{30}}$

Step 1 Write the numerator as a single fraction.
$$8+\dfrac{4}{x}=\dfrac{8}{1}+\dfrac{4}{x}=\dfrac{8x}{x}+\dfrac{4}{x}=\dfrac{8x+4}{x}$$
Do the same with each denominator.
$$\dfrac{x}{6}+\dfrac{1}{12}=\dfrac{x(2)}{6(2)}+\dfrac{1}{12}=\dfrac{2x}{12}+\dfrac{1}{12}=\dfrac{2x+1}{12}$$
Step 2 Write the equivalent complex fraction as a division problem.
$$\dfrac{\dfrac{8x+4}{x}}{\dfrac{2x+1}{12}}=\dfrac{8x+4}{x}\div\dfrac{2x+1}{12}$$
Step 3 Divide by multiplying by the reciprocal.
$$\dfrac{8x+4}{x}\div\dfrac{2x+1}{12}=\dfrac{8x+4}{x}\cdot\dfrac{12}{2x+1}$$
$$=\dfrac{4(2x+1)}{x}\cdot\dfrac{12}{2x+1}$$
$$=\dfrac{48}{x}$$

2. Simplify the complex fraction.   **2.** Simplify the complex fraction.

$\dfrac{\dfrac{rs^2}{t^3}}{\dfrac{s^3}{r^3t}}$ $\dfrac{\dfrac{a^3b^2}{c}}{\dfrac{a^5b}{c^3}}$

Use the definition of division and then the fundamental property.
$$=\dfrac{rs^2}{t^3}\div\dfrac{s^3}{r^3t}$$
$$=\dfrac{rs^2}{t^3}\cdot\dfrac{r^3t}{s^3}$$
$$=\dfrac{r^4}{st^2}$$

CHAPTER 4 Inverse, Exponential, and Logarithmic Functions

Name:
Instructor:

Date:
Section:

3. Simplify the complex fraction.

$$\frac{\dfrac{30}{x+4}-6}{\dfrac{4}{x+4}+1} = \frac{\dfrac{30}{x+4}-\dfrac{6(x+4)}{x+4}}{\dfrac{4}{x+4}+\dfrac{1(x+4)}{x+4}}$$

$$= \frac{\dfrac{30-6(x+4)}{x+4}}{\dfrac{4+1(x+4)}{x+4}}$$

$$= \frac{\dfrac{30-6x-24}{x+4}}{\dfrac{4+x+4}{x+4}}$$

$$= \frac{\dfrac{6-6x}{x+4}}{\dfrac{x+8}{x+4}}$$

$$= \frac{6-6x}{x+4} \cdot \frac{x+4}{x+8}$$

$$= \frac{6-6x}{x+8}$$

3. Simplify the complex fraction.

$$\frac{\dfrac{20}{x-5}-9}{\dfrac{5}{x-5}+2}$$

_____

Name:
Instructor:
Date:
Section:

4. Simplify each complex fraction.

a. $\dfrac{\dfrac{3}{10}+\dfrac{7}{20}}{\dfrac{1}{5}+\dfrac{2}{15}}$

Step 1 Find the LCD for all the denominators.
The LCD for 5, 15, 10, and 20 is 60.
Step 2 Multiply the numerator and denominator of the complex fraction by the LCD.

$$\dfrac{\dfrac{3}{10}+\dfrac{7}{20}}{\dfrac{1}{5}+\dfrac{2}{15}} = \dfrac{60\left(\dfrac{3}{10}+\dfrac{7}{20}\right)}{60\left(\dfrac{1}{5}+\dfrac{2}{15}\right)}$$

$$= \dfrac{60\left(\dfrac{3}{10}\right)+60\left(\dfrac{7}{20}\right)}{60\left(\dfrac{1}{5}\right)+60\left(\dfrac{2}{15}\right)}$$

$$= \dfrac{18+21}{12+8}$$

$$= \dfrac{39}{20}$$

b. $\dfrac{12+\dfrac{4}{x}}{\dfrac{x}{5}+\dfrac{1}{15}}$

Step 1 Find the LCD for all the denominators.
The LCD for x, 5, and 15 is $15x$.
Step 2 Multiply the numerator and denominator of the complex fraction by the LCD.

$$\dfrac{12+\dfrac{4}{x}}{\dfrac{x}{5}+\dfrac{1}{15}} = \dfrac{15x\left(12+\dfrac{4}{x}\right)}{15x\left(\dfrac{x}{5}+\dfrac{1}{15}\right)}$$

$$= \dfrac{180x+60}{3x^2+x}$$

$$= \dfrac{60(3x+1)}{x(3x+1)}$$

$$= \dfrac{60}{x}$$

4. Simplify each complex fraction.

a. $\dfrac{\dfrac{1}{3}+\dfrac{4}{9}}{\dfrac{5}{8}+\dfrac{3}{4}}$

_____

b. $\dfrac{4+\dfrac{2}{x}}{\dfrac{x}{3}+\dfrac{1}{6}}$

_____

Name: Date:

Instructor: Section:

5. Simplify the complex fraction.
$$\frac{\dfrac{9}{7n}-\dfrac{3}{n^2}}{\dfrac{8}{3n}+\dfrac{5}{6n^2}}$$

The LCD is $42n^2$.

$$=\frac{42n^2\left(\dfrac{9}{7n}-\dfrac{3}{n^2}\right)}{42n^2\left(\dfrac{8}{3n}+\dfrac{5}{6n^2}\right)}$$

$$=\frac{42n^2\left(\dfrac{9}{7n}\right)-42n^2\left(\dfrac{3}{n^2}\right)}{42n^2\left(\dfrac{8}{3n}\right)+42n^2\left(\dfrac{5}{6n^2}\right)}$$

$$=\frac{54n-126}{112n+35},\ \text{or}\ \frac{18(3n-7)}{7(16n+5)}$$

5. Simplify the complex fraction.
$$\frac{\dfrac{2}{9n}-\dfrac{2}{5n^2}}{\dfrac{4}{5n}+\dfrac{2}{3n^2}}$$

_____

6. Simplify each complex fraction. Use either method.

a. $\dfrac{\dfrac{6}{y}+\dfrac{5}{y+5}}{\dfrac{8}{y}-\dfrac{2}{y+5}}$

Method 2. Multiply by the LCD $y(y+5)$.

$$=\frac{\left(\dfrac{6}{y}+\dfrac{5}{y+5}\right)y(y+5)}{\left(\dfrac{8}{y}-\dfrac{2}{y+5}\right)y(y+5)}$$

$$=\frac{\left(\dfrac{6}{y}\right)y(y+5)+\left(\dfrac{5}{y+5}\right)y(y+5)}{\left(\dfrac{8}{y}\right)y(y+5)-\left(\dfrac{2}{y+5}\right)y(y+5)}$$

$$=\frac{6(y+5)+5y}{8(y+5)-2y}$$

$$=\frac{6y+30+5y}{8y+40-2y}$$

$$=\frac{11y+30}{6y+40}$$

6. Simplify each complex fraction. Use either method.

a. $\dfrac{\dfrac{2}{x}+\dfrac{3}{x+3}}{\dfrac{5}{x}-\dfrac{4}{x+3}}$

_____

Name: Date:
Instructor: Section:

b. $\dfrac{2-\dfrac{5}{x}-\dfrac{7}{x^2}}{1-\dfrac{3}{x}-\dfrac{4}{x^2}}$

Method 2. The LCD is x^2.

$= \dfrac{\left(2-\dfrac{5}{x}-\dfrac{7}{x^2}\right)x^2}{\left(1-\dfrac{3}{x}-\dfrac{4}{x^2}\right)x^2}$

$= \dfrac{2x^2-5x-7}{x^2-3x-4}$

$= \dfrac{(2x-7)(x+1)}{(x-4)(x+1)}$

$= \dfrac{2x-7}{x-4}$

c. $\dfrac{\dfrac{x+5}{x-4}}{\dfrac{x^2-25}{x^2-16}}$

Method 1.

$= \dfrac{x+5}{x-4} \div \dfrac{x^2-25}{x^2-16}$

$= \dfrac{x+5}{x-4} \cdot \dfrac{x^2-16}{x^2-25}$

$= \dfrac{(x+5)(x+4)(x-4)}{(x-4)(x+5)(x-5)}$

$= \dfrac{x+4}{x-5}$

b. $\dfrac{1-\dfrac{5}{x}+\dfrac{4}{x^2}}{1-\dfrac{7}{x}+\dfrac{12}{x^2}}$

_____

c. $\dfrac{\dfrac{3x+5}{x-2}}{\dfrac{9x^2-25}{x^2-4}}$

_____

Practice Problems

For extra help, see the videos on simplifying complex fractions in your MyMathLab course.

Simplify each complex fraction by writing it as a division problem.

1. $\dfrac{\dfrac{49m^3}{18n^5}}{\dfrac{21m}{27n^2}}$

1. _____

Copyright © 2015 Pearson Education, Inc.

Name: Date:

Instructor: Section:

2. Round 841 to the nearest hundred.

Step 1 Locate the place to which the number is to be rounded. Draw a line under that place.

   8̲41
   ↑ hundreds place

Step 2 Because the next digit to the right of the underlined place is 4, which is 4 or less, do not change the digit in the underlined place.

   ↓ Next digit is 4 or less.
   8̲41
   ↑ 8 remains 8.

Step 3 Change all the digits to the right of the underlined place to zeros.

   8̲41 rounded to the nearest hundred is 800.

In other words, 841 is closer to 800 than 900.

2. Round 520 to the nearest hundred.

_____

3. Round 86,949 to the nearest thousand.

Step 1 Locate the place to which the number is to be rounded. Draw a line under that place.

   86̲,949
     ↑ thousands place

Step 2 Because the next digit to the right of the underlined place is 9, which is 5 or more, add 1 to the underlined place.

    ↓ Next digit is 5 or more.
   86̲,949
     ↑ Change 6 to 7.

Step 3 Change all the digits to the right of the underlined place to zeros.

   86̲,949 rounded to the nearest thousand is 87,000.

In other words, 86,949 is closer to 87,000 than 86,000.

3. Round 24,866 to the nearest thousand.

_____

Name: Date:
Instructor: Section:

4. **a.** Round 46,989 to the nearest ten-thousand.

   Step 1 4̲6,989 (Ten-thousands place)
   Step 2 The next digit to the right is 6, which is 5 or more.
   ↓ Next digit is 5 or more.
   4̲6,989
   ↑ Change 4 to 5.
   Step 3 46̲,989 (Change 989 to 000.)
   46,989 rounded to the nearest ten-thousand is 47,000.

   b. Round 687,401,327 to the nearest million.

   Step 1 687̲,401,327 (Millions place)
   Step 2 The next digit to the right is 4, which is 4 or less.
   ↓ Next digit is 4 or less.
   687̲,401,327
   ↑ Leave 7 as 7.
   Step 3 687̲,401,327 (Change 401,327 to 000,000.)
   687,401,327 rounded to the nearest million is 687,000,000.

5. Round 3219 to the nearest ten, hundred, and thousand.

   First round 3219 to the nearest ten.
   ↓ Next digit is 5 or more.
   321̲9
   ↑ Tens place $(1+1=2)$
   3219 rounded to the nearest ten is 3220.

   Now go back to 3219, the original number, before rounding to the nearest hundred.
   ↓ Next digit is 4 or less.
   32̲19
   ↑ Hundreds place stays the same.
   3219 rounded to the nearest hundred is 3200.

   Again, go back to the original number before rounding to the nearest thousand.
   ↓ Next digit is 4 or less.
   3̲219
   ↑ Thousands place stays the same.
   3219 rounded to the nearest thousand is 3000.

4. **a.** Round 24,975 to the nearest ten-thousand.

   _____

   b. Round 836,502,671 to the nearest million.

   _____

5. Round 5467 to the nearest ten, hundred, and thousand.

   _____

Name: Date:

Instructor: Section:

6. Round 16.87453 to the nearest thousandth.

Step 1 Draw a "cut-off" line after the thousandths place.
  16.874 | 53

Step 2 Look only at the first digit you are cutting off. Ignore the other digits you are cutting off.
  16.874 | 53   (Ignore the 3)

Step 3 If the first digit you are cutting off is 5 or more, round up the part of the number you are keeping.
  16.874 | 53
  + 0.001
  ──────
  16.875

So, 16.87453 rounded to the nearest thousandth is 16.875. We can write $16.87453 \approx 16.875$.

7. Round to the place indicated.

a. 6.4387 to the nearest tenth

Step 1 Draw a cut-off line after the tenths place.
  6.4 | 387

Step 2 Look only at the 3.
  6.4 | 387   (Ignore the 8 and 7)

Step 3 The first digit is 4 or less, so the part you are keeping stays the same.
  6.4 | 387
  6.4

Rounding 6.4387 to the nearest tenth is 6.4. We can write $6.4387 \approx 6.4$.

6. Round 43.80290 to the nearest thousandth.

  ─────────────

7. Round to the place indicated.

a. 0.7976 to the nearest hundredth

  ─────────────

b. 0.79846 to the nearest hundredth

Step 1 Draw a cut-off line after the hundredths place.
  0.79 | 846

Step 2 Look only at the 8.
  0.79 | 846

Step 3 The first digit is 5 or more, so round up by adding 1 hundredth to the part you are keeping.

$$\begin{array}{r} \overset{1}{0.79} \mid 846 \\ +\ 0.01 \\ \hline 0.80 \end{array}$$

0.79846 rounded to the nearest hundredth is 0.80. We can write $0.79846 \approx 0.80$.

c. 0.02709 to the nearest thousandth

  0.027 | 09

The first digit cut is 4 or less, so the part you are keeping stays the same.
0.02709 rounded to the nearest thousandth is 0.027. We can write $0.02709 \approx 0.027$.

d. 64.983 to the nearest tenth

  64.9 | 83

The first digit cut is 5 or more, so round up by adding 1 tenth to the part you are keeping.

$$\begin{array}{r} \overset{1}{64.9} \mid 83 \\ +\ 0.1 \\ \hline 65.0 \end{array}$$

64.983 rounded to the nearest tenth is 65.0. We can write $64.983 \approx 65.0$. You must write the 0 in the tenths place to show that the number was rounded to the nearest tenth.

b. 7.7804 to the nearest hundredth

_____

c. 22.0397 to the nearest thousandth

_____

d. 0.649 to the nearest tenth

_____

CHAPTER 4 Inverse, Exponential, and Logarithmic Functions

Name: Date:

Instructor: Section:

Practice Problems

For extra help for exercises 1–11, see the videos on rounding in your MyMathLab course.

Locate the place to which the number is rounded by underlining the appropriate digit.

1. 257,301   Nearest ten   1. _____

2. 1037   Nearest hundred   2. _____

3. 645,371   Nearest ten-thousand   3. _____

Round each number as indicated.

4. 1382 to the nearest ten   4. _____

5. 16,968 to the nearest hundred   5. _____

6. 476,943 to the nearest ten-thousand   6. _____

Select the phrase that makes the sentence correct.

7. When rounding a number to the nearest tenth, if the digit in the hundredths place is 5 or more, round the digit in the tenths place (up/down).   7. _____

8. When rounding a number to the nearest hundredth, look at the digit in the (tenth/thousandth) place.   8. _____

Round each number to the place indicated.

9. 489.84 to the nearest tenth   9. _____

10. 54.4029 to the nearest hundredth   10. _____

11. 989.98982 to the nearest thousandth   11. _____

Name: Date:
Instructor: Section:

Chapter 5 Systems and Matrices

5.1 R Evaluating Expressions for Given Values; Inverse Property of Multiplication; Inverse Property of Addition

5.8 R Identity and Inverse Properties of Addition and Multiplication of Real Numbers

For review material about evaluating expressions for given value, see section 2.1R Guided Examples 1 and 2 and Practice Problems 1–3.

For review material about the inverse properties of multiplication and addition, see section 1.1R Guided Example 7 and Practice Problems 10–12.

For review material about the identity properties of multiplication and addition, see section 1.1R Guided Example 5 and Practice Problems 7–8.

Chapter 1 Equations and Inequalities

1.1R Properties of Real Numbers, Evaluating Exponential Expressions, Order of Operations, Simplifying Expressions, Operations on Real Numbers

Key Terms

1. identity element for addition
2. identity element for multiplication
3. exponential expression
4. base
5. exponent
6. numerical coefficient
7. term
8. like terms
9. sum
10. addends
11. minuend
12. subtrahend
13. difference
14. quotient
15. reciprocals
16. product

Now Try

1a. -12
1b. 2
2a. 4
2b. $[(-3) \cdot 4]$
3a. associative
3b. commutative
3c. both
4a. 107
4b. 12,600
5a. 0
5b. 1
6a. $\dfrac{7}{9}$
6b. $\dfrac{2}{3}$
7a. 11
7b. -8
7c. 0
7d. $\dfrac{5}{8}$
7e. -10
7f. 1
8a. 54
8b. $12y + 72 + 12x$
8c. $-13x - 78$
8d. $17x - 102$
8e. $30x + 84y + 126z$
8f. 375
8g. $23(x - y)$
9a. $-3x - 4$
9b. $10x + 7$
9c. $4x + 5y - z$
10a. 49
10b. 64
10c. 4096
10d. $\dfrac{25}{36}$

10e. 0.36 11a. 2 11b. 14

11c. $-\dfrac{5}{4}$ 12a. $-6m$ 12b. $23r$

12c. $19x$ 12d. $8x^2$ 12e. cannot be combined

13a. $49y+15$ 13b. $37k-24$ 13c. $-12+12r$

13d. $-\dfrac{5}{4}x+6$ 14. 6 15. -5

16. 3 17a. -12 17b. -31

17c. -40 18. -12 19. 2

20a. 3 20b. -4 20c. -38

20d. 2 20e. $\dfrac{61}{45}$ 21a. -56

21b. -3 21c. -45.08 22a. 30

22b. 135 22c. 18 22d. 90

23a. 6 23b. -6 23c. 0.31

23d. $\dfrac{36}{35}$ 24a. 3 24b. -2

24c. 3 24d. $\dfrac{4}{11}$

Practice Problems

1. 4 3. $(4+z)$ 5. $[(-4+3y)]$

7. 4 9. $\dfrac{6}{7}$ 11. 0; identity

13. $2an-4bn+6cn$ 15. $2k-7$ 17. $\dfrac{16}{81}$

19. 64 21. $\dfrac{16}{21}$ 23. unlike

25. $-10y^2+16y$ 27. $-1.5y+16$ 29. -18

31. $-5\dfrac{5}{8}$ 33. $\dfrac{1}{35}$ 35. -3

37. 46 39. $5\dfrac{7}{8}$ 41. $-\dfrac{2}{15}$

43. 40 45. 1.36 47. 0

1.2R Operations with Decimals; Translating Word Phrases into Algebraic Expressions and Equations; Formulas and Percent

Key Terms

1. factor
2. decimal places
3. product
4. equation
5. variable
6. algebraic expression
7. solution
8. constant
9. percent
10. percent equation

Now Try

1a. 15.630 1b. 21.709 2a. 36.844

2b. 1013.931 3a. 11.612 3b. 10.57

4a. 4.654 4b. 3.26 4c. 0.501

5. 10.793 6. 0.00186 7a. 1.413

7b. 15.03 8. 20.178 9. 16.791

10a. 20,410 10b. 1.78 11a. $x + 4$, or $4 + x$

11b. $10 - x$ 11c. $12 - x$ 11d. $20x$

11e. $\dfrac{10}{x}$ 11f. $5(x - 6)$ 12a. $3(x + 7) = 30$

12b. $5x + 3 = 50$ 12c. $3x - 7 = 12$ 13a. $63

13b. 91 packages 13c. 186 students 14a. 120 units

14b. 2491 points 14c. 60 15a. 29%

15b. 20% 15c. 320% 15d. 0.4%

IRWA-4 Integrated Review Worksheets Answers

Practice Problems

1. 92.49
3. 72.453
5. 42.566
7. 90.71
9. 0.0037
11. 4.271
13. 3.796
15. 162.791
17. $8x - 11$
19. $\dfrac{10}{x} = 2 + x$
21. $61 - 7x = 13 + x$
23. 106.4
25. 160
27. 22.9
29. 5%

1.3R Radical Notation; Simplifying Square Root Radicals; Rationalizing Square Root Denominators; Addition, Subtraction, and Multiplication of Square Root Radicals; Product Rule for Exponents; Zero Exponent Rule; Negative Exponent Rule; Add, Subtract, and Multiply Polynomials

Key Terms

1. radicand
2. perfect square
3. square root
4. radical expression
5. principal square root
6. irrational number
7. radical
8. rationalizing the denominator
9. unlike radicals
10. like radicals
11. product rule for exponents
12. degree of a term
13. descending powers
14. term
15. trinomial
16. polynomial
17. monomial
18. degree of a polynomial
19. binomial
20. like terms
21. inner product
22. FOIL
23. outer product

Now Try

1. 9, –9
2a. 13
2b. 41
2c. $\dfrac{3}{7}$
2d. 0.8
3a. 19
3b. 37
3c. $n^2 + 5$
4a. irrational
4b. rational
4c. not a real number
5a. 73

5b. 37   5c. $|n|$ 5d. $|n|$

6a. $2\sqrt{21}$ 6b. $9\sqrt{2}$ 6c. cannot be simplified

7a. $10y\sqrt{y}$ 7b. $4m^2r^4\sqrt{3mr}$ 8a. $\dfrac{2\sqrt{15}}{15}$

8b. $\dfrac{3\sqrt{42}}{7}$ 8c. $-\dfrac{\sqrt{3}}{5}$ 9a. $-\dfrac{3\sqrt{10}}{8}$

9b. $\dfrac{9x\sqrt{2xt}}{t^3}$, $x \geq 0$, $t > 0$ 10a. $-\sqrt{6}$

10b. $13\sqrt{2z}$ 10c. cannot be simplified

10d. $\dfrac{2\sqrt{5}}{3}$ 11a. $\sqrt{10}-\sqrt{55}+\sqrt{6}-\sqrt{33}$ 11b. $6+3\sqrt{7}-2\sqrt{2}-\sqrt{14}$

11c. 12   11d. $11-6\sqrt{2}$ 11e. $16-\sqrt[3]{4}$

11f. $p-q$, $p \geq 0, q \geq 0$ 12a. 9^{13} 12b. $(-6)^6$

12c. x^6 12d. m^{27} 12e. 500

12f. 108   13a. 1   13b. 1

13c. -1 13d. 1   13e. 88

13f. 1   14a. $\dfrac{1}{64}$ 14b. $\dfrac{1}{27}$

14c. 25   14d. $\dfrac{49}{36}$ 14e. $\dfrac{8}{27}$

14f. $\dfrac{1}{8}$ 14g. $\dfrac{11}{p^4}$ 14h. x^8

14i. $\dfrac{b^4}{a^5}$ 15a. $\dfrac{125}{36}$ 15b. $\dfrac{y}{x^7}$

15c. $\dfrac{qr^5}{4p^3}$ 15d. $\dfrac{625b^4}{a^4}$ 16a. $5x^3-5x^2+4x$

16b. $9x^4+7x^2+7x-6$ 17a. $4x^3+x+12$ 17b. $7x^3+8x^2-14x-1$

18a. $2x+2$ 18b. $-11x^3-7x+1$ 19. $9x^3+x^2-11x+7$

20. $7+8x-4x^2$ 21a. $13a+ab$ 21b. x^2y+4xy

22a. $32x^4+64x^3$ 22b. $-35m^8+42m^7-28m^6+7m^5$

23. $4x^7-2x^5+37x^4-18x^2+9x$

24. $28x^4-33x^3+51x^2+17x-15$ 25. $x^2+3x-54$

26. $16xy+72y-14x-63$ 27a. $15k^2+44kn+32n^2$

27b. $45p^2+11pq-4q^2$ 27c. $35x^5-195x^4-90x^3$

Practice Problems

1. $25, -25$ 3. $\dfrac{30}{7}$ 5. not a real number

7. $2xy^3z^5\sqrt{2xz}$ 9. $\dfrac{y\sqrt{21b}}{6b}$ 11. $\sqrt{10}-4\sqrt{5}+2\sqrt{3}-4\sqrt{6}$

13. 7^7 15. $48k^{18}$ 17. -2

19. $-\dfrac{2}{k^4}$ 21. $\dfrac{2y^7}{3x^4}$ 23. $4x^2+6x-18$

25. $-8w^3+21w^2-15$ 27. $3x^3+7x^2+2$ 29. $-5ab-6ac$

31. $35z^4+14z$ 33. $-6y^5-9y^4+12y^3-33y^2$

35. $6m^5+4m^4-5m^3+2m^2-4m$ 37. $20a^2+11ab-3b^2$

39. $-6m^2-mn+12n^2$

1.4R, 1.5R   Factoring Out the Greatest Common Factor; Factoring Trinomials; Factoring Binomials

Key Terms

1. factoring 2. factored form 3. greatest common factor

4. factor 5. prime polynomial 6. difference

7. perfect square trinomial

Now Try

1a. $8x(x^4+3)$
1b. $4y^2(5y^2-3y+1)$
1c. $n^7(n+1)$

1d. $8x^3(8x^2-5x+1)$
2. $-9pq(pq^2+3p^3q-1)$
3a. $(y+8)(y+4)$

3b. $(z+5)(z^2-11)$
4. $(x+3)(x+8)$
5. $(y-7)(y-5)$

6. $(p+9)(p-3)$
7. $(a-17)(a+2)$
8a. prime

8b. prime
9. $(p-7q)(p+2q)$
10. $7x^4(x-5)(x-2)$

11a. $(z+6)(z-6)$
11b. $(r+s)(r-s)$
11c. prime

11d. prime
12a. $(2x+9)(2x-9)$
12b. $(5t+7)(5t-7)$

12c. $(8r+3s)(8r-3s)$
13a. $10(3x+7)(3x-7)$
13b. $(x^2+8)(x^2-8)$

13c. $(p^2+16)(p+4)(p-4)$
14. $(p+8)^2$

15a. $(x-12)^2$
15b. $(8m+3)^2$
15c. prime

15d. $5x(2x+5)^2$
16a. $(t-6)(t^2+6t+36)$
16b. $(3k-y)(9k^2+3ky+y^2)$

16c. $3(x-4)(x^2+4x+16)$
17a. $(6x+1)(36x^2-6x+1)$

17b. $6(x+2y)(x^2-2xy+4y^2)$

Practice Problems

1. $10x(2x+4xy-7y^2)$
3. $-13x^8(-2+x^4-4x^2)$
5. $(x-7)(x-4)$

7. $2n^2(n-5)(n-3)$
9. $10k^4(k+2)(k+5)$
11. $(9x^2+4)(3x+2)(3x-2)$

13. $\left(z-\frac{2}{3}\right)^2$
15. $-3(2a-5b)^2$
17. $8(3x-y)(9x^2+3xy+y^2)$

19. $(3r+2s)(9r^2-6rs+4s^2)$

21. $(4x+7y)(16x^2-28xy+49y^2)$

1.6R Rational Expressions; Lowest Terms of a Rational Expression; Operations with Rational Expressions; Factoring (including by substitution); Negative and Rational Exponents; Simplifying Radicals with Index Greater than 2

Key Terms

1. rational function
2. rational expression
3. standard form
4. quadratic in form
5. power rule for exponents
6. product rule for exponents
7. quotient rule for exponents
8. index (order)
9. cube root
10. index, radicand

Now Try

1a. $2; \{x \mid x \neq 2\}$
1b. $2, 3; \{x \mid x \neq 2, 3\}$

1c. none; $\{x \mid x \text{ is a real number}\}$

1d. none; $\{x \mid x \text{ is a real number}\}$
2a. $\dfrac{m}{3}$

2b. already in lowest terms
2c. $\dfrac{3y+2}{2y+1}$

2d. $\dfrac{x-3}{4}$
2e. $x^2 + x + 1$
2f. $\dfrac{3x+2}{4x-3}$

3a. -1
3b. $-(x+10)$, or $-x-10$
4a. $\dfrac{5y+7z}{11}$

4b. $-\dfrac{2}{z^2}$
4c. $\dfrac{1}{c+d}$
4d. $\dfrac{1}{x-1}$

5a. $\dfrac{19}{18z}$
5b. $\dfrac{s-30}{s(s-6)}$
6a. 3

6b. $\dfrac{-30}{(x+3)(x-3)}$
7. $\dfrac{x-12}{x-8}$
8. $\dfrac{1}{x}$

9. $\dfrac{8x^2+37x+15}{(x+5)(x-5)^2}$
10. $\dfrac{9p-15}{(p-5)(p-1)(p+1)}$
11a. $\dfrac{36x}{7}$

11b. $\dfrac{m}{m-2}$
11c. $\dfrac{4}{5}$
11d. $\dfrac{(x-1)(x+4)}{x(x+2)}$

11e. $\dfrac{1}{x-3}$
12a. $\dfrac{2}{9p}$
12b. $\dfrac{5x}{3}$

12c. $\dfrac{2a+3}{a+3}$
13a. $u^2 - 5u + 4 - 0$
13b. $u^2 + 2u - 35 - 0$

13c. $u^2 - 2u - 3 = 0$ 14a. $\{-2, -1, 1, 2\}$ 14b. $\left\{-1, -\frac{3}{4}, \frac{3}{4}, 1\right\}$

14c. $\left\{-\sqrt{2-2\sqrt{3}}, -\sqrt{2+2\sqrt{3}}, \sqrt{2-2\sqrt{3}}, \sqrt{2+2\sqrt{3}}\right\}$ 15a. $\frac{1}{64}$

15b. $\frac{1}{27}$ 15c. 25 15d. $\frac{49}{36}$

15e. $\frac{8}{27}$ 15f. $\frac{1}{8}$ 15g. $\frac{11}{p^4}$

15h. x^8 15i. $\frac{b^4}{a^5}$ 16a. $\frac{125}{36}$

16b. $\frac{y}{x^7}$ 16c. $\frac{qr^5}{4p^3}$ 16d. $\frac{625b^4}{a^4}$

17a. 6 17b. 11 17c. −2

17d. not a real number 17e. −3 17f. $\frac{1}{2}$

18a. 9 18b. 8 18c. −216

18d. 9 18e. not a real number 19a. $\frac{1}{4}$

19b. $\frac{1}{625}$ 19c. $\frac{16}{9}$ 20a. $\sqrt[3]{23}$

20b. $\left(\sqrt[4]{21}\right)^3$ 20c. $5\left(\sqrt[4]{x}\right)^5$ 20d. $\left(\sqrt[3]{2x}\right)^4 - 3\left(\sqrt[5]{x}\right)^2$

20e. $\frac{1}{\left(\sqrt{x}\right)^3}$ 20f. $\sqrt[4]{x^2 - y^2}$ 21a. $22^{1/2}$

21b. $10^2 = 100$ 21c. m 22a. $5^{5/2}$

22b. $a^{2/15}$ 22c. $x^2 y^{7/2}$ 22d. $\frac{x^2}{y^{1/4}}$

22e. $r^{7/6} - r^{19/6}$ 23a. $x^{7/6}$ 23b. $y^{7/12}$

23c. $x^{1/4}$ 24a. 5 24b. −5

24c. 7 25a. 5 25b. −5

25c. not a real number 25d. −5 25e. −5

26a. 4   26b. −5   26c. −2

26d. $-x^4$ 26e. w^{10} 26f. $|x^5|$

27a. $\dfrac{\sqrt[3]{10}}{5}$ 27b. $\dfrac{\sqrt[4]{4tx^3}}{x}$ 28a. $\sqrt[3]{21}$

28b. $\sqrt[3]{35xy}$ 28c. $\sqrt[5]{8w^4}$

28d. cannot be simplified using the product rule directly

28e. $16 - \sqrt[3]{4}$

Practice Problems

1. $\dfrac{2}{3}$; $\left\{s \mid s \neq \dfrac{2}{3}\right\}$ 3. 1, 2; $\{q \mid q \neq 1, 2\}$ 5. $\dfrac{y+1}{y-3}$

7. $\dfrac{4n-7}{m+3}$ 9. $\dfrac{1}{k-4}$ 11. $\dfrac{6z^2 + 19z - 4}{(z+2)(z+3)(z+4)}$

13. $\dfrac{x-3}{x+2}$ 15. $\dfrac{m-2}{m-4}$ 17. $\dfrac{-3(6k-1)}{3k-1}$

19. $\left\{-\dfrac{1}{2}, \dfrac{1}{2}, -\sqrt{5}, \sqrt{5}\right\}$ 21. $\dfrac{1}{m^{18}n^9}$

23. −4   25. −15   27. 7776

29. $4\left(\sqrt[5]{y}\right)^2 + \sqrt[5]{5x}$ 31. $a^{1/4}$, or $\sqrt[4]{a}$ 33. $a^{5/6}$

35. −4   37. −1   39. $-x$

41. $\sqrt[3]{x^2}$ 43. $\dfrac{\sqrt[3]{35x^2}}{7x}$ 45. $\sqrt[8]{28}$

47. $4 - \sqrt[3]{25}$

1.7R Sets and Set Operations; Order on the Number Lines
Key Terms

1. negative number   2. positive number   3. union

4. compound inequality   5. intersection

Now Try

1. true
2. ∅
3. [–2, 8]
4. [–5, 0)
5. ∅
6. {–6,–5,–4,–3,–2,–1}
7. $(-\infty,-2) \cup [2, \infty)$
8. $[7, \infty)$
9. $(-\infty, \infty)$

Practice Problems

1. true
3. false
5. {2, 4}
7. [5, 9]
9. ∅
11. {1, 2, 3, 4, 5, 6, 7, 8, 9, 10}
13. $(-\infty,-4] \cup (4, \infty)$

1.8R Definition and Properties of Absolute Value; Evaluating Absolute Value Expressions

Key Terms

1. absolute value

Now Try

1a. 10
1b. 10
1c. –10
1d. –10
1e. 3
1f. –3
1g. –3

Practice Problems

1. –10
3. 2

Chapter 2 Graphs and Functions

2.1R Evaluating Expressions for Given Values

Key Terms

1. variable
2. algebraic expression
3. constant

IRWA-12 Integrated Review Worksheets Answers

Now Try

1a. 24, 36 1b. 252, 567 2a. 64

2b. 20 2c. 80

Practice Problems

1. 8 3. $\dfrac{28}{13}$

2.2R Squaring a Binomial; Factoring Perfect Square Trinomials

Key Terms

1. binomial 2. perfect square trinomial

Now Try

1. $x^2 + 12x + 36$ 2a. $64a^2 - 48ab + 9b^2$ 2b. $4a^2 + 36ak + 81k^2$

2c. $9a^2 - 66ab + 121b^2$ 2d. $9p^2 + p + \dfrac{1}{36}$ 2e. $45x^3 - 210x^2 + 245x$

Practice Problems

1. $49 + 14x + x^2$ 3. $16y^2 - 5.6y + 0.49$

2.3R Solving for a Specified Variable

Key Terms

1. formula

Now Try

1a. $W = 5.5$ 1b. $b = 31$ 3. $t = \dfrac{d}{r}$

4. $a = P - b - c$ 5. $h = \dfrac{2A}{b+B}$ 6a. $y = 6x - 18$

6b. $y = 6x + 5$

Practice Problems

1. $a = 36$ 3. $h = 12$

5. $n = \dfrac{S}{180} + 2$ or $n = \dfrac{S + 360}{180}$

Integrated Review Worksheets Answers **IRWA-13**

2.4R Division Involving Zero

Key Terms

1. remainder
2. dividend
3. quotient
4. divisor

Now Try

1a. 0
1b. 0
1c. 0
1d. 0
2a. undefined
2b. undefined
2c. undefined

Practice Problems

1. $a = 36$
3. $h = 12$
5. $n = \dfrac{S}{180} + 2$ or $n = \dfrac{S+360}{180}$

2.8R Operations with Polynomials; Operations with Rational Expressions

Key Terms

1. dividend
2. quotient
3. divisor

Now Try

1. $x^2 - 3x$
2. $3n^2 - 4n - \dfrac{2}{n}$
3. $-4z^3 + \dfrac{7}{2}z^4 - \dfrac{5}{z} - \dfrac{3}{z^2}$

Practice Problems

1. $2a^3 - 3a$
3. $-13m^2 + 4m - \dfrac{5}{m^2}$

Chapter 3 Polynomial and Rational Functions

3.2R Dividing Polynomials

Key Terms

1. divisor
2. quotient
3. dividend

Now Try

1. $-2a^3b^2 - 4a^2b + 3$
2. $2x^2 - 2x - 2 + \dfrac{-5}{5x-1}$
3. $x^2 + 10x + 100$
4. $3x^2 + 5x + 5 + \dfrac{8x+29}{x^2-4}$
5. $x^2 + \dfrac{1}{8}x + \dfrac{5}{8} + \dfrac{6}{8x-8}$

Copyright © 2015 Pearson Education, Inc.

Practice Problems

1. $-3x + 4$
3. $a^2 + 1$

3.6R Using Geometric Formulas; Ratio and Proportion

Key Terms

1. vertical angles
2. formula
3. perimeter
4. area
5. ratio
6. numerator; denominator
7. proportion
8. cross products

Now Try

1. 12 ft
2. 15 ft, 20 ft, 30 ft
3. 25 m

4a. 148°, 148°
4b. 54°, 126°

5a. $\dfrac{17}{8}$
5b. $\dfrac{8}{5}$
6a. $\dfrac{2}{3}$

6b. $\dfrac{2}{1}$
6c. $\dfrac{2}{3}$
7. $\dfrac{3}{8}$

8a. $\dfrac{19}{9}$
8b. $\dfrac{22}{17}$
9. $\dfrac{5}{7}$

10a. $\dfrac{24}{17} = \dfrac{72}{51}$
10b. $\dfrac{\$10}{7 \text{ cans}} = \dfrac{\$60}{42 \text{ cans}}$
11a. $\dfrac{9}{7} \neq \dfrac{4}{3}$, false

11b. $\dfrac{1}{3} = \dfrac{1}{3}$, true
12a. $306 = 306$, true
12b. $32 \neq 35$, false

13a. 32
13b. 10.67
14a. $2\dfrac{3}{5}$

14b. 6.4

Practice Problems

1. 8 in.
3. 31,400 sq ft
5. 59°, 121°
7. $\dfrac{25}{19}$
9. $\dfrac{1}{4}$
11. $\dfrac{9}{2}$
13. $\dfrac{2}{7}$
15. $\dfrac{5}{6}$
17. $\dfrac{36}{45} = \dfrac{8}{10}$
19. $\dfrac{4}{3} = \dfrac{3}{4}$; false
21. $\dfrac{7}{2} = \dfrac{4}{1}$; false
23. $165\dfrac{3}{5} = 165\dfrac{3}{5}$; true
25. 36
27. 40
29. 21

Chapter 4   INVERSE, EXPONENTIAL, AND LOGARITHMIC FUNCTIONS

4.1R   Simplifying Complex Fractions

Key Terms

1. complex fraction
2. LCD

Now Try

1a. $\dfrac{35}{51}$
1b. $\dfrac{90}{x}$
2. $\dfrac{bc^2}{a^2}$
3. $\dfrac{-9x+65}{2x-5}$
4a. $\dfrac{56}{99}$
4b. $\dfrac{12}{x}$
5. $\dfrac{5n-9}{3(6n+5)}$
6a. $\dfrac{5x+6}{x+15}$
6b. $\dfrac{x-1}{x-3}$
6c. $\dfrac{x+2}{3x-5}$

Practice Problems

1. $\dfrac{7m^2}{2n^3}$
3. $\dfrac{9s+12}{6s^2+2s}$ or $\dfrac{3(3s+4)}{2s(3s+1)}$
5. $\dfrac{(x-2)^2}{x(x+2)}$

4.5R   Rounding Whole Numbers and Decimals

Key Terms

1. rounding
2. estimate
3. decimal places

Now Try

1a. 8<u>6</u>4 is closer to 860.
1b. <u>2</u>398 is closer to 2000.

1c. 9<u>3</u>7,645 is closer to 940,000.
2. 500

3. 25,000
4a. 20,000
4b. 837,000,000

5. 5470, 5500, 5000
6. 43.803
7a. 0.80

7b. 7.78
7c. 22.040
7d. 0.6

Practice Problems

1. 257,3<u>0</u>1
3. 6<u>4</u>5,371
5. 17,000

7. up
9. 489.8
11. 989.990